T0324806

Pricing and Sustainability of Urban Real Estate

Urban sustainability has become a political and social agenda of global significance, of which real estate is an integral dimension. Sustainable urban development includes much more than 'green building' standards, yet in practice, other aspects such land use plans and locations are often overlooked.

This book demonstrates that the issue of sustainable development stretches far beyond the hitherto dominating agenda based on 'green' (i.e. environmentally and ecologically sustainable) buildings. In doing so, it presents a novel framework based on the concept of economic sustainability of real estate locations, drawing connections with the global financial crisis and housing price bubble discourse. It argues for the need to better integrate social, cultural and economic dimensions into the real estate sustainability agenda. It also explores the role of location, and especially the image aspect therein. Trends in consumer choice are important to the way these dimensions are appreciated in decisions about investment, development, valuation and other activities of the production, consumption and governance of the built environment.

This book will be of interest to private and public sector practitioners of real estate valuation as well as scholars of urban studies, geography, economics, urban planning and environmental studies.

Tom Kauko is Senior Lecturer in Real Estate at the School of the Built Environment, Oxford Brookes University, UK.

Pricing and Sustainability of Urban Real Estate

Tom Kauko

Routledge
Taylor & Francis Group

LONDON AND NEW YORK

First published 2017
by Routledge
2 Park Square, Milton Park, Abingdon, Oxon OX14 4RN

and by Routledge
711 Third Avenue, New York, NY 10017

Routledge is an imprint of the Taylor & Francis Group, an informa business

© 2017 Tom Kauko

British Library Cataloguing in Publication Data
A catalogue record for this book is available from the British Library

Library of Congress Cataloging in Publication Data
Names: Kauko, Tom, author.
Title: Pricing and sustainability of urban real estate / Tom Kauko.
Description: Abingdon, Oxon ; New York, NY : Routledge, 2017. | Includes bibliographical references.
Identifiers: LCCN 2016038492 | ISBN 9781472462435 (hardback) | ISBN 9781315602110 (ebook)
Subjects: LCSH: Real estate development—Environmental aspects. | Sustainable urban development. | Real estate investment. | Real property—Valuation.
Classification: LCC HD1390 .K38 2017 | DDC 333.3309173/2—dc23
LC record available at https://lccn.loc.gov/2016038492

ISBN: 978-1-4724-6243-5 (hbk)
ISBN: 978-1-315-60211-0 (ebk)

Typeset in Times New Roman
by Apex CoVantage, LLC

Contents

Figures

Tables

Foreword

During the past decade, significant progress has been made in the area of sustainable design, construction and management of real estate assets. A largely shared and relatively sound understanding of what sustainability means in relation to both single buildings and organisations has been developed within the construction and real estate industry. On the basis of this understanding, systems, tools and metrics for the sustainability assessment of buildings as well as for corporate real estate sustainability management and reporting have emerged. Today, the issue of sustainability is increasingly being seen as an organisational success factor and within many construction and real estate firms efforts are being undertaken to embed sustainability considerations within regular decision-making processes and information flows at different hierarchical levels.

In addition, professional bodies and governmental authorities have developed guidance, guidelines and standards for various built environment professionals on how to take sustainability considerations into account within everyday practice. Furthermore, environmental legislation with regard to the construction and real estate sector is becoming increasingly stringent; particularly within European Union member states. In combination with increased levels of consumer awareness regarding environmental concerns, all this has stipulated the discussion about the relationship between sustainability related features of buildings and economic aspects like market prices and rents.

Empirical research carried out during the past couple of years has shown that market participants' preferences in favour of sustainable buildings' credentials already do impact on the price formation process: within some sub-markets and in some countries property pricing is increasingly distinguishing between buildings that exhibit different sustainability related building features and associated physical or operational performance. There is recognition that buildings which are not resource efficient, low carbon in terms of operation and location and which are not equipped to flex to changing occupier needs will not be future proofed in market value terms. And this, in return, impacts on value stability and likely value development of all properties in the marketplace. In this context, the argument is often made that further recognition and integration of sustainability considerations in property valuation standards and valuation and risk assessment practices is a key driver for further market transformation towards more sustainable real estate markets.

However, this discussion and the developments described earlier very often take place with the implicit or explicit focus on single buildings and their immediate surroundings. On the one hand, the question whether or not real estate market developments are actually sustainable from a socio-economic point of view (e.g. real estate affordability and diversity, social cohesion) is not often addressed within the sustainable building literature. On the other hand, quantitative housing market analysis and modelling and sustainable development concerns are traditionally treated as incompatible.

Nonetheless, a truly sustainable real estate market is not only about the sustainability related features of single buildings and about sustainability conscious companies and professionals. It is also about social and economic longevity and viability. And that is the topic of the present book. Hitherto neglected issues are covered in this book.

Within eight interconnected essays Tom Kauko discusses and defines appropriate criteria and metrics for judging the socio-economic sustainability of real estate markets and provides examples for practical applications in terms of empirical analyses of selected markets. In doing so, he bridges two gaps: first, he contributes to shifting the prevailing perspective of the sustainable building community well beyond the single building and its surrounding. Second, he contributes to embedding sustainability concerns into housing market analysis and modelling.

Both are necessary for truly informed decision making and for further transformation towards truly sustainable real estate markets.

David Lorenz
Professor for Property Valuation and Sustainability
Centre for Real Estate, Karlsruhe Institute of Technology (KIT)

Preface

This publication comprises a collection of rather independent essays. It ties together my more than fifteen years of research at the interface of *urban sustainability* and *sustainable real estate* research areas. While it incorporates a sample of current, more general discourses relating to urban economic geography and housing market analysis, the aim here is first and foremost to look at possibilities to broaden the sustainability discourse in the direction of measurable indicators, in particular, property prices. I have already long preached about the massive amount of information that price data contains, and when combined with other kinds of data on social, environmental and institutional issues, even better. In this endeavour it furthermore becomes relevant to investigate the viability of alternative approaches too. Given all this, the focus of this compilation, while varying between the chapters, stays mostly on socio-economic dimensions, spatial elements and alternative research methodology. This is where the originality of this lies.

This work has no ideological stance but to some extent fits into world views of a realist or pragmatist research approach. This is because I believe that the world is too fragmented for the Grande, almost religion-like theoretical approaches of the not-too-distant past. Sustainability research is of course a relatively new thing and, while much is allowed, the approach taken here is open to debate too. However, if you criticise, you must have an alternative. So in this realm one needs to be constructive – at least this is my take on the situation.

As for the opening photo, it is from a front-garden of a single-family house in a Hungarian small town (actually my wife's family's home). After a bathtub change, instead of disposing of the old one it was just left here (presumably a combination of cost saving and sheer laziness in the face of all complications of transporting the bathtub away). However, the wind blew a rich pile of seeds there and together with rainwater and dirt this sedimentation begun to yield organic growth within a few months. The sheer randomness of the process is astonishing! And believe it or not: this pic is by no means from the first summer of flourishing! I would say that the bathtub-flowerpot kind of gives an innovative edge to an already beautiful garden. Talk about synergy – in this case, saving work and adding to the aesthetic and biologic side – a topic that runs throughout this book. It is hard to think of a more illustrative example of complete bottom-up organic sustainability processes at a micro-locational scale; the bloom being a symbol of spontaneous market development and dynamics!

This argumentation has benefited from various informal sources, notably following the panel sessions in 'mega' events such as the EXPO REAL Munich (October 2008) and the EUSEW Brussels (I have visited this event most years since 2007); discussions with Dr. David Lorenz, Lorenz Property Advisors and University of Karlsruhe (in various meetings since October 2008); and the participants of the following conferences: EUGEO Amsterdam, August 2007; RSA Prague, May 2008; EUGEO Bratislava, August 2009; RSA Leuven, 6–8 April 2009; ERSA Liverpool, 27–31 August 2008; and ENHR Prague, June 2009, the housing finance workshop, session on residential property valuation. You are too many to list here, but I thank you all for earnest, constructive and 'non-nonsense' input. I must furthermore thank the suppliers of the data: the data for the Trondheim part was originally acquired from Norsk Eiendomsinformasjon AS and subsequently generously supplied by NTNU and the project

VULCLIM; furthermore, the respective sources for the other data are the tax office (Gemeentebelastingen) of Amsterdam and the Hungarian Statistical Office (KSH). I also like to thank Katy Crossan and Amanda Buxton formerly at Ashgate for the final analysis concerning the suitability for publication of this manuscript. Obviously I am indebted to many others for their helpful suggestions and clarifications – you know who you are, and credit where the credit is due.

Tom Kauko

1 The sustainable urban real estate agenda

There is more to it than the 'green' thing

Introduction

Sustainable cities have existed much longer than the sustainability[1] discourse. From a historical perspective it can be argued that private investment was always paramount for creating the great cities where ingenuity flourished in both civic and business arenas. This was the case in late medieval Florence and (perhaps incorrectly) is assumed to be the case in contemporary Manhattan – the 'Florence of our times'. Innovations in the banking-sector tend to generate property investments and eventually the accumulated prosperity further fosters high-level social and cultural life. In other words, sustainable investment practices much steer urban development – historically and in today's circumstances, where 'sustainability' runs the risk of becoming a buzzword. Therefore, when assessing various possibilities for such sustainable investment practices, one vital task we cannot escape is to assess the contribution to energy savings that could be made by cities, their buildings and their citizens. To leave it here would, however, only serve only a narrowly defined sustainability agenda that neglects dimensions other than the 'green' ones. In fact, it can be argued that social and economic sustainability issues show more resonance with property investors than that of the older type of environmental-energy sustainability (see Sayce et al., 2007).

This study merges and meshes concepts from a variety of specific urban and real estate discourses. In doing so it argues for the prime importance of an economic sustainability dimension rather than the environmental and social dimensions. Ideally, it would be the other way, but in today's reality this is unfortunately not the case. The basic assumption here is that we need economic growth, to be able to extract the funds necessary to deliver other, 'softer' amenities. There really is no other way. Otherwise, the population growth – which is happening anyway – will cause scarcity of resources on such a level that conflicts become unavoidable. This goes on as long as the majority of people on this globe are not either altruistic or long-term planners. Exceptions obviously exist, such as those who have too much and those who have nothing and do not know of materially better circumstances, but it would be naïve to assume that such marginal actors have a substantial role yet in today's development and market making of urban real estate.

The sustainability discourse stretches, of course, beyond investment issues that drive urban development. It is evident that progress in sustainable urban development requires improvement in our capacities of acting professionally as well as in the civic sphere. How can academics contribute to such improvement? The idea is to sell what resonates with the masses rather than to sell the science per se; while this 'product' needs to be based in solid science, what matters is the possibility to improve decision-making communities related to the urban environment. It is furthermore not only about the poorest on the globe, but it is also challenging to sort the problems of those who were part of a 'western middle-class', but who subsequently have fallen due to unemployment, mortgage default or other personal crisis. Obviously, the key question is as to whose responsibilities fixing these problems are: the politicians', the NGO's or the private sector's?

In the same spirit as the point about Florence earlier, Keivani (2009) underscores the importance of understanding interactions between environmental concerns and social and economic domains (this triple bottom line approach to sustainability is nevertheless different from the stance taken in the present study). The economic role of cities has now gained further importance due to the current era of globalisation, he continues. The issue is also one of accessibility; namely, policy making can be influenced and development agendas can be best set in cities as these are centres of political power and administration, Keivani adds. He also notes that social issues are often more contrasted in cities than elsewhere – and this is obviously an alarming development in the global south. Indeed, cities pose huge challenges for us but fortunately, due to agglomeration economies/benefits, cities also allow great efficiency in the use of resources as well as offer space for greater innovation and production (cf. Glaeser, 2011).[2]

Regardless of one's particular strand taken, there is an apparent need for a more precise conceptualisation of sustainable development.[3] While various competing visions of sustainable urbanism exist, these comprise a multiplicity of potential futures. The problem is that in some areas of knowledge the advances have been huge and in others less so. The challenge is that while the notion of the sustainable city is immediately appealing, it also is complex and intangible. Here a 'one-size-fits-all' model is inadequate to understand the way social and technical solutions interact; in fact, these two spheres of action (i.e. the social and technical realms) need to be combined into an integrated approach. Some solutions require top-down actions whereas others require more of a bottom-up approach. It is furthermore important to admit that despite planning for change, not all changes can be foreseen. It is realistic nevertheless to see some opportunities arising in the midst of the crisis. (Williams, 2010; cf. Fisher, 2010).

The problem of top-down and bottom-up coordination together with the problem of dealing with uncertainty and opportunities arising involve various spatial scales. At the extremes we might compare for instance an individual house-buyer at a given time and place with the institutional investor's systematic pooling of portfolios. And sustainable development is argued to be spatial, although the scale varies (Zuindeau, 2006). Real estate is a spatial factor of sustainability too. The spatial element is in fact one of the core issues of this book. And as real estate is a vital ingredient in the urban environment it becomes paramount for the argumentation here: how to use real estate data to design a socio-economic sustainability approach based on measurable indicators (e.g. house price).

Indeed, according to the United Nations Economic Commission for Europe (UNECE) real estate is important for the economy, especially in relation to the financial crisis. Furthermore, a sustainable real estate market, which would be reached by appropriate regulations and education of actors, would therefore enhance the sustainability of the economy. UNECE has stated ten principles for reaching a sustainable real estate market: integrated legal framework, efficient land register and cadastre, efficiency of services, prerequisites for development of sound real estate markets, good governance, sustainable financing, transparency and advanced financial products, property valuation, social housing, training and capacity building. This issue involves measures to be implemented at the local level and in cooperation with the private sector. In particular, an inter-sectorial approach is required here (UN, 2010).

While the abovementioned logical argument supports an overall strengthening of the real estate economy, the reality is different. Traditionally, buildings or their locations are not perceived as a core operation in one's business, neither from the occupiers' nor the investors' point of view. However, this perception is now changing, as the long-term occupancy costs *de facto* affect the way the business is run (e.g. McMillen, 2005; see also Addae-Dapaah et al., 2009). Thus it can be argued that sustainability matters in real estate context. On the other hand, without definitions of criteria any arguments concerning sustainability remain vague. Therefore, on the basis of various literatures ranging from economics and social sciences to planning and construction engineering, I have compiled the following list of twelve issues that currently surround the urban land use and real estate sustainability agenda. These are listed from the most localised (the building) to the widest (region) scale.[4]

Some of the issues are selected for further discussion, insofar as they are central for the argumentation of this contribution.

1   Energy efficiency in buildings (during their life cycles)

When assessing various possibilities for such sustainable investment practices as discussed so far, one vital task we cannot escape is to assess the contribution to energy savings that could be made by cities, their buildings and their citizens. EU objectives aim at 20% energy savings by 2020.

This figure of energy saving in buildings can for example be achieved by an appropriately-designed lighting system which reduces total building electricity (Jackson, 2009, p. 96). Zalejska-Jonsson (2013), however, does not detect a major impact of energy and environmental building on the purchasing or renting decision.

2   Use of renewable energy in buildings (during their life cycles)
3   Pollution control in building (during their life cycles)

EU objectives aim at 20% reduction in greenhouse gas emissions by 2010.

4   Real estate quality (economic sustainability)

As a grossly substandard level of housing is unacceptable for health and safety reasons, the quality (largely a subjective indicator though) ought to develop in the same direction and with the same pace as the price level. This concerns the site and building specific attributes as well as the characteristics of the surrounding environment, neighbourhood and the city as a whole. Related to this argument, Jones et al. (2009) suggest that the affluence of the area – thus the socio-economic situation and its consequence, the neighbourhood house price level – is a more important determinant for the viability of a housing development than planning policy or risk and costs caused by brownfield developments. In other words, the sustainability of new housing development varies across local housing market areas and also within cities, and that this is not so much due to the compact cities policy as it is due to the affluence of the neighbourhood where the development takes place, according to this study. For such economic sustainability arguments to be taken on board by investors and developers as well as architects and regulators above all requires a steady and sustained price lift, and not only for the price levels set by the project but for the property values created in the surrounding area and adjacent neighbourhoods too. In this way, prices and values of the dwellings and rents of office space contain powerful information for the purpose of deciding on the feasibility of the strategy from an economic sustainability point of view. What unfortunately complicates the picture is that while the property value in fact can be substantially increased by acknowledging sustainability in general and the environmental-ecological dimension (i.e. 'green' issues) in particular, it can also be increased by unsustainable elements (Kauko, 2011b).

5   Real estate affordability (economic sustainability)

As it is not sufficient with high quality unless people cannot afford to buy (or rent) it, the affordability (often approximated as net income) of the dwelling also ought to develop in the same direction and with the same pace as the price level.

6   Real estate diversity (economic sustainability, with causal links to points 10 and 12)

Even if the quality and affordability criteria are fulfilled, it is not sufficient for economic sustainability unless there is a wide enough range (i.e. product variety generated for most apt selections to be made) of different quality and affordability levels on the market. This is because the drivers of sustainability: production technology, community governance and consumption fashions all tend to change quickly and then it is vital not to have neglected any specific property/housing package even if it may seem marginal at some stage. Or put it differently, if a potential market trend setter or other innovation in terms of quality or affordability is not recognised this will have harmful impacts for the evolution of the property portfolio (cf. Foxon et al., 2012).

The role of diversity has (following Polèse and Stren) become an established concept in the international urban sustainability discussion and is relevant also here. However, not all diversity is sustainable – it can be seen either as 'an asset and an engine of the sustainable development of the city' or, less politically correctly, 'as a liability and a source of potential tension and conflict' whenever the cultural differences between the natives and the newcomers become too wide for their peaceful coexistence (Bitušiková and Luther, 2010).[5] In general, management of diversity in an urban area (public space together with land and housing) influences its social and cultural sustainability in two hypothetical directions: increased diversity either is sustainable (theory) or unsustainable (practice of failures of certain immigrant groups to integrate in western European and North American cities).

7   Optimal density for a block/neighbourhood (with causal links to points 10 and 12)

Compact city is often applied as a model for a sustainable city but in many cases, and for numerous reasons, this model has failed; for example, because sealed surfaces without green spaces are vulnerable to climate effects. In economic terms higher density normally should be considered more energy efficient, but there are exceptions; for example, small projects might allow better use of innovations, as they can be tailored to the particular green features, such as water collection systems (e.g. Ganser and Williams, 2007), or a heating-cooling system that utilises differences between day and night time ground temperatures (see e.g. REEEP, 2011). Or it can be similarly concluded that the highest building efficiencies, while environmentally sustainable, are not socially sustainable due to the deteriorated sense of community and safety experienced by the inhabitants (see e.g. Bramley et al., 2009). The literature here is abundant. Reading this literature the conclusion is that the optimal density is relatively high but not as high as possible.

8   Public transportation availability (functional issue)

Lucas and colleagues (2010) in turn emphasise the role of accessibility planning. They argue that sustainable communities require sufficient poor people's accessibility, which will be achieved by improving public transport together with land use decisions concerning accessibility. They illustrate this by the case of Thames Waterways, where the travel possibilities with public transportation are currently considered poorly planned.

9   Traffic pollution (ecological issue)
10 Social cohesion in the neighbourhood/city/region (with causal link to point 12)

Social sustainability has been approached by Manzi et al. (2010a, b), whose starting point is to define sustainability in terms of social equity, access to resources, participation, social capital, human rights and exclusion. The nature of social sustainability is holistic embracing inclusion, care and governance (p. 10). The key question posed by Manzi and colleagues (2010b) concerns the competence of governments to steer partnerships and networks so as to achieve social sustainability in the sense of incorporating a wider range of stakeholders in the delivery of urban processes (pp. 10–15). Here is obviously a link to point 7.

According to Manzi et al. (2010b, p. 17), sustainable communities are defined by eight sectors (the *Egan Wheel model*): governance, transport and connectivity, services, environment, equity, economy, housing and the built environment, and social and cultural factors. Moreover, while sustainability agendas [*sic*] neglect the political dimensions, evidence tells us that social sustainability levels are highest in egalitarian societies (p. 22). Manzi and colleagues (2010b) conclude that, despite conceptual flaws and practical difficulties, social sustainability is an important guiding principle for direction of policies and environment; furthermore, these authors argue that it is inseparable in relation to environmental and economic dimensions. The overall conclusion of Manzi and colleagues (2010a) is that social sustainability requires investment and not only plans.

To open up a different avenue of research, White and colleagues (2010) argue that the impact of ICT on reaching social sustainability is far from straightforward and may even be counter-intuitive. The crux of the argument here is that the travel reduction mainly concerns high income workers, and therefore ICT is likely to lead to a more divisive development than what one would intuitively assume. This argumentation ties with the *Smart City* debate. According to the proponents of the *Smart City*, hi-tech solutions that utilise electronic data collected through sensors and card readers would help planning for people's movements and making life more comfortable as well as improving the security concerns. Examples of this type of development include Songdo in South Korea and Masdar in United Arab Emirates. (We come back to the *Smart City* approach and these two places in Chapter 3.)

This overt reliance of ICT also exposes this approach to critique from mainly leftist and environmentalist circles; it can namely be argued that democracy is lacking in a place so completely dominated by propagation of technology and surveillance systems. Instead an alternative approach based on a more community oriented agenda is proposed (one well-known example being the Holzmarkt project in Berlin).

11 Communicativeness in local/regional planning (governance transparency)
12 Innovativeness of the region (economic sustainability, including financial transparency of corporations, and also favouring local products and labour, see Figure 1.1.)

Following an evolutionary economic (and more recently, also a complexity theoretical as well as an experimental) approach, it can be argued that sustainable development of urban land use and real estate situations is to a great extent related to innovation – including social innovations. The key to understanding the economic sustainability concept is to realise how incentives are set up to stimulate work, saving and reinvestment. Economic sustainability is, however, not a new concept. Already in 17th century Europe a distinction could be made between the sustainable Spanish kingdom and the unsustainable Polish ones. In the former investments were made in public services for the people; for example, the first street lights were developed

Figure 1.1 The point of this brochure is to show that the availability of local food and drink products represents a specific type of economic sustainability.

here. In the latter kingdom the revenues were just consumed by an elite group of 5% of the population whereas the rest of the population was starving. We could make similar notions of how unsustainable regimes communism produced, and in contemporary times about the EU banking crisis which also is 'disincentivising' for the ordinary taxpayer.

Bryson and Lombardi (2009) purport that maximising short-term profits in residential development projects in cities leads to unsustainable development in the long run. Conversely, if normal profits are reaped and the remaining margins (i.e. the extra profits) are reinvested wisely the longevity of the project – an important precondition for sustainability – can be enhanced. Thus the argument concerns how the extra profits from real estate and built environment related long-term investments could be used to generate sustainability in other than economic terms too. For example, the project developer invests in bus stops, in more green features of the buildings and other areas or even in some innovations that improve the social and environmental sustainability of the dwellings and residential areas. Here the key to finance non-economic elements is to achieve long-term profit margins from property development activity. How to get private developers interested in this, however, will not be easy, as long-term economic strategies are required from them. This in turn requires good governance and designing apt institutions to direct the investments on the right track. In other words: economic sustainability must come before the social aspects discussed earlier and the more standardised/established environmental aspects.

As already implied, for any sustainable development to occur the ultimate challenge is to prepare for a long time-horizon – say, at least two generations ahead.[6] If we accept an evolutionary perspective to sustainability – like RICS recently does, see for example Macintosh (2010) and Ratcliffe et al. (2010) – the key to success is innovativeness. This requires heterogeneity in product ranges, which in turn is fostered by flexible and market sensible administrative structures and is influenced by the changing tastes of individuals as consumers and citizens. That the differences across product packages become increasingly qualitative also means an accentuated role for the vicinity and the environment in the analysis. Here is also a link to point 6 in the list.

In practice many of the twelve points listed previously are interlinked. The EU sustainability strategy for example focuses on climate change and clean energy, sustainable transport, consumption and production, conservation and management of natural resources, public health, social inclusion, demography and migration as well as global poverty. Currently, in the US, in turn, sustainability is defined in the context of the 'environmental justice agenda' according to which marginal and poor groups should not disproportionately bear the costs of public or private activities or policies.

While precise causal relationships are uncertain, policies to ameliorate the negative environmental and socio-economic externalities are necessary to ensure sustainable communities. (Manzi et al., 2010a, b).

This volume is organised as a series of interconnected essays that each pick one or more of the abovementioned issues (and if more issues, at times meshing them together). The essays do furthermore relate to each other in a loose sense only. The reason for this relative patchiness of the text is due to the observation that sustainable real estate analysis can pertain to so many particular issues and agendas. As an academic research objective, sustainable buildings and areas are currently approached from three different literatures:

1   The impact of the building on their users – health issues of the workers and residents in particular (micro approaches);
2   Global impacts (macro approaches): emissions, energy efficiency and renewable energy;
3   Urban and environmental sustainability of cities/city-regions (meso approaches).

The focus here is on the third tradition: more precisely, about development of areas including both brownfield and greenfield sites. Zuindeau (2006) brings up intergenerational equity – the long-term criteria for sustainable development – and notes that a similar definition for sustainability differences across space and different territories is yet to be developed. He subsequently introduces a spatial measure of sustainability: namely, ecological footprints in relation to the available bio-capacity: that is ecological deficit or self-sufficiency. Both measures – density and bio-capacity – can be related to financial compensation to form a monetary measure of sustainable development. This tie to the main arguments – the backbone – purported in this book: that sustainability includes much more issues than the 'green building' standards so commonly approximated with 'sustainable building' features.

Here it is to note that this argumentation necessarily disagrees with authorities such as Du Plessis and Cole (2011), who argue for shifting the sustainability paradigm and moving the sustainability debate towards a more eco-centric one; here it is argued that economy must come first, otherwise the problem will be to get funds to improve the environment.[7] Indeed, it is agreeable that we need a paradigm shift, but one that is concrete and geared towards socio-economic issues rather than what is proposed by Du Plessis and Cole. Besides, the fact is that the green dimension is already well explored. Was not there already too much emphasis on environmental sustainability in comparison with social, cultural and economic sustainability? Ever since the Brundtland report of the late 1980s, the eco-logic and environmental issues have been at the fore here. At least this is one of the assumptions underlying this study.

Thus, the corollary is that the sustainable real estate also includes, at least in principle, socio-economic issues. And for credibility's sake these are advised to be measurable indicators. One such indicator may of course be the monetary value of the abovementioned 'green features' – most commonly isolated using a hedonic regression approach. Furthermore spatial and alternative approaches (i.e. to the linear ones) are worth investigating in this realm. The overall goal of this kind of approach to real estate sustainability is to create added value for sustainable real estate analysis.

In this volume the socio-economic, spatial issues relate to three elements in a sustainability context: buildings, locations and land use plans. Each of these elements is examined as follows.

Buildings

The sustainability of residential, retail and office buildings is much about selecting the apt production technology. Sustainability is per definition a local issue, for example an individual is encouraged to consume and produce locally as much as possible; however, the problems in relation to research scarcity, urban inequality, financial crisis and so forth are of a global character. As will be discussed in Chapter 3, there is a recently emerging empirical tradition within neoclassical economics that is preoccupied with testing the hypothesis of 'green' buildings having a price premium or other economic benefit compared to a standard (or unsustainable) building. It is also possible that the popularity of green buildings over their non-green counterparts has or has not to do with the attitudes and beliefs of the occupants towards environmental behaviour, as evidence from Malaysia suggests (see Wilkinson et al., 2013).

Locations

As for the residential location and surroundings of the dwelling, the theoretical starting point concerns environmental-ecologic, social-cultural and economic-financial dimensions. Ideally all three criteria are required for total sustainability, but in practice it is about tradeoffs and a largely elusive knowledge base with non-standard definitions, as Støa (2009) has pointed out. For example, an increased building efficiency (i.e. floor-space per land area; density) is sustainable environmentally and economically but often unsustainable socially; in fact, in Britain Bramley and colleagues (2009) and Bramley and Power (2009) have confirmed this empirically, which then would contradict the conventional wisdom of higher density leading to community cohesion. Furthermore, adding the process perspective to the static perspective, when we look at how to manage these issues (i.e. behavioural aspect), what is referred to as 'good community governance' needs the support of the private sector too. If looked at from a more top-down angle this means smart policies, regulations and especially incentives set at the local and regional levels, as such institutional and policy tools would be an imperative to meet the sustainability goals set out in the Rio-1992 agenda (see Lützkendorf et al., 2011). The last thing that comes to mind is that one needs to be prepared for catastrophes such as flooding (or as in the Norwegian city of Trondheim, quick clay landslide) when designing the building and its location.

Land use plans

When examining the role of land use plans, the starting point is to realise the harsh realities: unsustainability poses massive problems (e.g. urban inequality and disinvestment) and these cannot be solved by design only. This problem is especially serious in metropolitan areas of North America and this is due to the lack of political will on one hand and due to preferences of households and businesses for low-density developments on the other hand (e.g. Reese and Sands, 2007). While the zoning apparatus (and planning in general) is somewhat overvalued in this context, planning issues such as governance and ethics are important when examining the goals of project and strategic development (see Bogliotti and Spangenberg, 2006). The planning apparatus includes relevant concepts and principles for steering the development such as smart growth and thereby comprises a secondary set of factors to complement those related to investment (e.g. Stewart et al., 2006; see also Coyle, 2011; Talen, 2011).

This is, however, not to deny a crucial (albeit subordinate) role for the multitude of laws, regulations, policies and practices that constitute the planning system. Planning is entered as a parameter into the market-based models in four ways (e.g. Adams et al., 2005):

1   Positive/proactive: to stimulate the markets
2   Neutral: to facilitate the markets
3   Negative: to regulate the market
4   Capacity building (most modern role): how to make the governance structures more efficient, effective and responsive by including private, public and civil society actors.

The zoning applies to the first three ways. The zoning plan can (1) encourage new building, or (2) merely facilitate such, or then (3) restrict such, for example with growth barriers and density caps. In the third case brownfield investment and refurbishment of existing dwellings become the available options. This study does not explicitly look at planning and policy contexts, however; rather, it looks at value creation and affordability in market settings.

Sustainable urban areas: what is so new about them?

The chain of argumentation could also begin from what really is 'new' in the 'sustainable urban area' debate compared to earlier literatures on similar things. To give some examples of the latter, spatial analysis of local economic and non-economic (i.e. aesthetic, cultural, social, ecological) values/benefits could be considered as such 'new things' in this context. Until *c.* early 2000s those literatures merely covered neighbourhood satisfaction/quality/effect, location value/preference/quality, place attachment/marketing, area attractiveness and similar problem areas that relate to an externality effect that would be measurable in a cost-benefit framework. Volumes of theoretical and empirical literatures within economics, psychology, sociology, geography and planning (notably, house price and residential preference models) much agree on their main conclusions: often enough residents tend to choose a place mainly based on social factors. As a consequence, in each case the most gains to be made are in understanding the social aspects (see e.g. Manzi et al., 2010b; Ahmed, 2011). In fact, the traditional humanistic argument, for example following Tuan, is that residents choose a place based on two largely contradictory criteria: on one hand, people want safety, familiarity and avoidance of nuisance; on the other, they want grandeur, excitement and stimuli at the same time (see also Kauko, 2004a).[8]

Within the urban sustainability paradigm the drivers of innovation are of three kinds: technology development, cultural branding and consumption issues and institutions, notably political leadership (Joss, 2011). While we undoubtedly possess plenty of knowledge of urban amenity and nuisance effects based on research, with the advent of sustainable development and urban sustainability the issue is about how to apply this knowledge in practice: to select the best starting point, elements and issues to prioritise and so forth. Today we have ·possibilities offered by technology based strategies (hi-tech or low-tech) on one hand, and cultural and arts strategies (e.g. street-art) on the other, and we can combine these too as well (e.g. smart art). There should therefore really be no obstacles to development.

A related, but far more pressing, issue concerns the economic realities of a given project. The financial stakeholders have a key role as to finance sustainable built environment (Lützkendorf et al., 2011). Who pays for the sustainability element, if not the property developer and investor? We need then to convince also representatives of these actors to join on-board to the participation procedures. In other words, while the normative element is ideally determined by the user (e.g. Leishman and Warren, 2010), the aim is to market the idea also to the producers of new technology and artefacts, not only to the consumers. Especially, at the time of financial downturn including sustainability features might not be economically viable – at least not in large-scale developments. That is why we also should look at small-scale and refurbishment agendas. The bottom line here is that in good times a developer driven agenda should do the trick, but in not-so-good times the urban economy needs a long-term 'smartish' and spatially differentiated strategy, where the abovementioned issues are factored in, with standards and protocols that ensure the outcome.

Urban governance and planning strategies here have a great challenge in the public delivery of services while also combating other issues such as land use sprawl, pollution and social segregation. It has been suggested that a future welfare system should ensure universal citizenship rights, recognise difference and be governed in a socially innovative way, thereby taking distance from the traditional bureaucratic, authoritarian one that was indifferent to diversity and needs. Instead, however, the current shift in delivery has reflected an odd convergence of neo-liberal policy preferences and bottom-up initiatives. Among the more successful cases is the *Pilestredet Park project* in Oslo in 2000 where a former hospital site was transformed into an eco-friendly residential area in the inner city (EC, 2010).

The urban real estate paradigm

The concepts brought up so far work on many levels and with a long time perspective. At least two different spatial levels are crucial for observing the contemporary issues related to urban economic sustainability: local and regional (possibly national, supranational and continental levels too, given what was already noted about the global side of the sustainability discourse). In both cases we must look at the long-term economic benefit and the convergence of the people and corporate decision makers (including social/ethical values). Furthermore, to promote diversity is argued to be more appropriate than a one best way forward, as variety leads to selection of the fittest (the most adaptable technology, policy or consumer fashion). Here some of the most important issues concern density and land use (see e.g. Reese and Sands, 2007). In the following we examine multifaceted concepts and long-term effects within this paradigm.

Multi-level concepts

Hill and colleagues (2011) conclude that built environment professionalism ought to take issues related to sustainable development more seriously – they request more emphasis on the 'why' rather than only covering the 'what' question as hitherto when dealing with the relationship between knowledge, skills and ethics to models of professional practice (see also Dent and Dalton, 2010; Warren-Myers, 2012). The broader issue here is about how to connect the sustainability gains of the real estate economy to the greening of the economy at large. The built environment represents the largest single contributor to environmental degradation and pollution emissions and offers the most cost-effective opportunities to contribute to a greener economy, by cutting down energy and resource use and improving human health at the same time (e.g. Lorenz et al., 2008). In the words of Hoornweg et al. (2011, p. 18), ' [. . .] what you buy is important, but what type of housing and neighbourhood you live in is much more important.' To name but a few key figures, 40% of world's greenhouse gases are caused by activities of the built environment life cycle (construction, management, maintenance, refurbishment and demolishing) and another 35% is related to the provision of supporting infrastructure (see Davoudi et al., 2008); furthermore, 70% of the pre-credit crunch global economy is said to pertain to indirect or direct ownership of real estate assets. From the point of view of the overall argument of greening the economy, it is inevitable that this sector too needs communication with and education of consumers and professions as well as financial incentives and regulations (see Lützkendorf et al., 2011).

At a global level, some 'local/regional politicians, technicians and residents in the process of area based regeneration' are pessimistic about current sustainability policies. Especially worried they are in relation to the duration and impact of the financial crisis. According to this view, the current economic policy is based on denial. Based on the observations that only 10% of the variation in quality-of-life (QOL) is explained by income or possessions, and that the bio-capacity of the Earth is insufficient for current growth strategies, their proposition is to 'reinvent growth itself' and to create 'a less materialistic society' (LUDEN, 2012).

Long-term effects

While the point here is about incremental changes rather than 'one grand planning vision', the further issue concerns how the economic sustainability can generate environmental and social sustainability by reinvesting the profits made with view on long-term developments (see Bryson and Lombardi, 2009). It needs to be stressed that for any sustainable development to occur the ultimate challenge is to prepare for a long time-horizon; say, at least to the next generation, but preferably much longer. This is one of the reasons why Hegedűs (2011), based on Hungarian evidence, finds gated community type developments to be unsustainable socially (as a rule, because of their segregation function) and also economically (dependent upon the quality level, type and timing of the buildings). If we accept an evolutionary perspective to sustainability, the key to success is innovativeness. This requires heterogeneity in product ranges, which in turn is fostered by flexible and market sensitive administrative structures and is influenced by the changing tastes of individuals – consumers and citizens. Therefore it can be asserted that real time management is far more important and effective than plans and traditional government bureaucracies. At the same time, however, a market-based agenda alone cannot lead us onto the right sustainable development track. Thus it can be argued that good governance and incentives as well as education including scenarios and forecasting are necessary to bring into this equation.

It is to observe that economic sustainability is not only about the (nth year) cost savings, but how one reinvests these savings so as to maximise the provision of public amenities, at least one generation (but preferable several generations) ahead. The actors involved include bureaucrats, professions, corporations and smaller market actors as well as citizens/residents.

Overview of the book

The provision of the built environment comprises real estate and infrastructure. This book focuses on the former element. The argument concerns how the extra profits from real estate and built environment related long-term investments could be used to generate sustainability in other than economic terms too. For example, the project developer invests in bus stops, in more green features of the buildings and other areas or even in some innovations that enhance the social and environmental sustainability of the urban residents (see Wagner et al., 2007). Here the key to finance non-economic elements is to achieve long-term profit margins from property development activity (following Bryson and Lombardi earlier). How to get private developers interested in this, however, will not be easy, as it requires long-term economic strategies from them. This in turn requires good governance and designing apt institutions to direct the investments on the right track. In other words: economic sustainability must come first – only then we can afford the relative luxury of considering green issues as well as the social issues outlined earlier.

At this early stage of the development of a paradigm, a coherent body of knowledge with unanimous evaluators does not exist yet. Different viewpoints and approaches are of course allowed, given the short history of the sustainability discourse – while the concept of sustainable development was used as early as 1972 in the Club of Rome report *Limits to Growth*, the first serious academic debates about its definitions emerged in the 1980s. Despite improvements still some of this critique remains today (cf. Støa,

2009; Manzi et al., 2010a, b; Colantonio and Dixon, 2011; Talen, 2011; Young and Dhanda, 2013).[9] Thus most of the issues at stake in the sustainability debate can hardly be considered completely recent phenomena. Following the particular view taken here, in an urban context progress can be argued to signify a paradigm change based on long-term administrative behaviour (via an institutional approach), long-term market behaviour (heterodox economics approach) and human behaviour in actors' consumption and location choices (behavioural approach), where economic sustainability is the key.

For economic sustainability arguments to be taken on board by developers as well as regulators above all requires a steady and sustained price lift, and not only for the price levels set by the project but for the property values created in the surrounding area and adjacent neighbourhoods too. Prices and values of the dwellings and rents of office space contain powerful information for the purpose of deciding on the feasibility of the strategy from an economic sustainability point of view. We can furthermore say that property values and valuations also have a double function as sustainability indicators: the property value in fact can be substantially increased by acknowledging sustainability in general and the environmental-ecological dimension (i.e. green issues) in particular, but that, unfortunately, it can also be increased by unsustainable elements (Kauko, 2011b).

When dealing with this relationship between price and sustainability indicators, the main solution championed is the improvement of valuation methods, and mainstreaming of automated valuation models (AVMs) in particular. The logic here is that improving and updating the estimates can be used as a basis for each new round of sound investments. We need, however, some quantitative as well as qualitative controls of the AVM output insofar as this concerns (1) smoothing temporary peaks and lows as well as (2) adding an ethical dimension to the valuations. How this is to be defined then is dependent on the particular sustainability context. AVMs need plenty of good quality data, however. The logic here is that, if AVMs enable better long-term investments, they then also would contribute to an enhancement of green and social issues. Without sound evidence base, decisions remain *ad hoc* or ideological and thereby contestable by definition. In particular, the risk is that emerging economies carry out the same mistakes of unsustainability as we have done during the last fifty years or so.

A propagation of a change in attitudes and procedures that direct real estate investment, management and development activity – a paradigm shift – would concern the criteria of decision making, which is not anymore about economic efficiency or liberalising the economy, but about long-term goals that are articulated in terms of sustainable development. This not only occurs in state-of-the-art academic research activity, but increasingly also in supranational level politics and policy making, as these bodies are committed to an agenda of sustainability including social cohesion and economic competitiveness at regional and local levels. The primary aim of this compilation of eight connected essays is to show that the issue of sustainable development stretches far beyond the hitherto dominating agenda based on 'green' (i.e. environmentally and ecologically sustainable) buildings. In doing so, a novel framework based on the concept of *economic sustainability* of real estate locations is elaborated. This also connects with the financial crisis, credit crunch and housing bubble discourse – it is to note that, traditionally, housing market modelling and sustainable development paradigms are treated as incompatible. For residential property three more specific issues are pertinent: (1) the price in relation to quality, (2) the price in relation to income (affordability), and (3) the diversity of the product. The empirical parts of the study concern the development of house prices in relation to incomes in Trondheim, Norway (in the period 1993–2007); the classification of the development of house prices in relation to subjective quality in various parts of Amsterdam, the Netherlands (during 1986–2002); and case studies of urban renewal areas in Budapest, Hungary (1998–2002), and Amsterdam respectively. The findings are suggested to have direct significance for sustainability assessment in a real estate context. In particular, urban residential segments are picked for scrutiny here.

As for the particular methodological approach taken, the aim is to investigate enhanced sustainability of urban amenity factors or procedures, including those relating to the green building dimension (cf. Wilkinson et al., 2013). This will be carried out using quantitative datasets of house sales price – an informative indicator of both sustainable and unsustainable elements. Expert interviews are used to support these results as well as to generate new ones – as well as different issues. My case studies will concern three very different European geographical contexts: Trondheim, Amsterdam and Budapest. The more quantitative part of the analysis is carried out using multidimensional clustering and classification algorithms (SOM, LVQ), and a method of target and comparative cases. The study is concluded by comparing the results of the literature review and empirical evidence gathered on various spatial scales.

This study discusses the definitions, diagnostics and implementation of economic sustainability metrics within an urban real estate context. For practical reasons, the empirical material pertains to the residential sector: data on house prices are simply more readily available than data on other types of properties. The chosen cases are all rather atypical insofar as none of them represents the Anglo-American circumstances most common in the literature. Such an atypical selection of cases will have certain advantages. First of all, it shows the nuances in the existing body of theory. Ideally, it also might help in adding to the literature, and thereby contributing to the originality of the study. In relatively new research areas such as real estate sustainability these are important benefits.

Last, while there are at first sight other sectors with perhaps more obvious connections to fuelling the economy, the centrality of the real estate sector should not be forgotten due to the vital functional/physical and financial role this sector plays. On one hand the real estate industry is providing the facilities for the companies who are active in these other industries; on the other hand real estate assets in themselves pertain to a substantial part of our globalised economy (cf. D'Arcy and Keogh, 1998). At the building and neighbourhood level the issues then is about providing real estate services that are also sustainable in various dimensions (see Wagner et al., 2007). The solution to this lies in designing an economic long-term strategy; the issue at stake is about making the buildings and neighbourhoods more sustainable by reinvesting the profits made from developments. When that is carried out consistently over the built environment of a whole urban area we are able to confirm patterns of greening of the economy. One feasible methodology to analyse the extent of these issues is to use house price data together with other local information.

Notes

1 In much of the urban literature sustainability is defined in a narrow sense as environmental sustainability only (i.e. the *green* aspect comprising CO_2 emissions, energy efficiency and renewables). In this study however the definition is broader and incorporates also social and economic sustainability dimensions, even if these are much less explored and understood than their green counterparts. In fact, the emphasis here is on the economic dimension.

2 However, agglomeration economies are hugely overvalued according to Turok (2004), who emphasises broader economic forces and policies (see also Budd and Hirmis, 2004).

3 Sustainability can be defined in many ways. The first serious academic debates about the difference in sustainability definitions emerged already in the 1980s. For example, Shanmugaratnam (1990) laments that the concept of sustainable development has multiple meanings, is in the danger of becoming overused and suffers from a too narrow definition of economic accounting systems. Despite improvements still some of this critique is aired today (cf. Støa, 2009; Manzi et al., 2010a, b; Colantonio and Dixon, 2011; Talen, 2011). Thus most of the issues at stake in the sustainability debate can hardly be considered as completely recent phenomena.

4 The list is by no means exhaustive; nonetheless, these points might have a certain appeal to anyone who analyses urban affairs, land use or area development in a sustainability framework.

5 As I am writing this, Huseby riots – an internationally noted conflict situation that lasted for several days in a Stockholm suburban housing estate populated mostly by immigrants – have just passed.

6 Informal discussion with Indy Johar, Director of the UK based consultancy firm Architecture 00 (12 April 2011).

7 And if cost saving is seen as an option here, we then again move towards a position that emphasises the economic side.

8 Population change might also be an issue that affects sustainability of a residential area, albeit here the logic behind does not follow what one would think intuitively. Namely, using an analytically simplistic endogenous econometric equilibrium model, Varvarigos and Zakaria (2011) show that not only does an increased population lead to congestion and subsequently lower environmental quality, but also the causality holds in reverse: improved environmental quality leads to higher longevity and also to a lower fertility rate. Thus adaption of cleaner production technology hampers the population growth.

9 For an altogether different kind of criticism, see Cook and Swyngedouw (2014), who encourage looking for 'utopian ideas' instead of what they call 'the sustainability industry' where [*sic*] socio-ecological justice and inequality is ignored. According to Cook and Swyngedouw, true urban sustainability should be political – not just technological and organisational. Their view would combine political ecology and environmental justice approaches. On the other hand, while such a view is praised among critical geographers, is it really constructive? In the present book the perspective offered is rather realist and pragmatic rather than utopian.

2 Economic sustainability considerations in a local real estate context (with post-bubble hindsight)

The magnitude and pace of the financial 'meltdown' in the world economy took us by surprise when it started towards the end of year 2007. At the end of the summer 2008 it was mainly about crisis in the US and then the UK, but two months later it had already reached global proportions. For instance, in Norway – the richest country in the world – the currency was devalued 10% and, while in this country the financial institutions were not in serious troubles, many Norwegian municipalities lost the total worth of their investments. In other countries outside the Euro-zone the problems were more serious, and the tone of government policy makers was hopeless. On the other hand, speaking about the US, the leader of the world economy, a homeowner confidence survey by *Zillow* found out that 62% of American homeowners believed that the value of their house had increased or stayed the same during a year when in reality 77% of US homes actually declined in value (*AVM News*, 2008a)! Apparently consumers tend to overvalue their assets with it-happens-to-others-but-not-to-me mentality. It is reasonable to believe that similar attitudes prevail on this side of 'the pond' too.

But is it really 'over' or is it just a new 'beginning'? Is the development of value and market modelling tools an endeavour worth embarking on, or are there unforeseeable risk factors in sight? Has room now, thanks to the economic downturn, been created for merging the 'sustainable development' and 'market modelling' paradigms? According to an old Finnish proverb 'there is nothing that bad that there cannot be something good in it'. Perhaps the new, more difficult situation forces us to sharpen our tools and see what can be made as efficiently as possible. It is furthermore acknowledged that it might be possible to cease this unique opportunity of market calmness to 'take stock' and learn from the financial crisis.[1] In other words, we have now all the right reasons to look at long-term growth patterns and evaluate how sustainable these are – and to begin speaking the same language as those engaged in sustainability research.[2] This forward looking amidst general awareness of mistakes made is taken from the discourse of the practitioners rather than that of the real estate and housing market research community, as the latter, due to academic inertia, can be considered a laggard in this respect (see Kauko, 2004c; Kauko and d'Amato, 2008).

On the other hand, the sustainable real estate research community has arguably not covered the socio-economic dimensions to a sufficient extent, even though monetary value estimates of green features such as eco-certification of buildings has already been produced using sophisticated hedonic modelling studies (as will be shown in Chapter 3 of this volume). In line with the overall argument of this compilation, the sustainability requirement goes far beyond the green issues, so as to include the cultural, social and economic conditions for real estate market and development. Given that this book explicitly champions a sustainability metrics, the first of these three dimensions might be too difficult to incorporate to the proposed approach. The latter two dimensions, however, should be feasible to deal with now that socio-economic info is stored in several databases, both private and public ones. The present chapter thus evaluates the extent to which such an approach – preferably spatial, and possibly somewhat alternative one – is likely to improve the real estate analysis in terms of efficiency, accuracy or some other criteria.

Within this apparent insight in mind, it is fair to say that the current theory of measuring house prices and mapping the markets is inadequate. This concerns, on the one hand, the price setting in relation to the development of affordability, and on the other hand, the price setting in relation to more static price factors tied to fundamental quality variables including location specific amenities such as the physical and social environment or transport infrastructure. While this topic is important in general socio-economic terms, as witnessed through the established urban housing market research tradition (see e.g. Maclennan, 1977; Ball and Kirwan, 1977; Maclennan and Tu, 1996; Watkins, 2001; see also Meen and Meen, 2003, for a somewhat alternative approach) we may also now speculate about a more particular issue: given the pivotal role of the housing bubble in the subsequent credit crunch, could the global financial crisis have been moderated or even avoided if prudent property valuations had been made at the time of issuing bank loans, and if governments and other institutions were concerned about raising the quality of property market information? Do *automatic valuation models/methods* (AVMs, automated, often computerised, procedures for carrying out the task of valuing one or more properties) help us in mortgage lending problems? And if we don't believe they do, what are the valid grounds for criticism? Here we may select between two different lines of analysis: we may, based on empirical evidence, argue for or against a direct empirical modelling application for mortgage valuation; alternatively, we may look at a more qualitative, rhetorical way of convincing either for or against the general use of AVM in this research area.

It is unlikely that we could come up with a methodology to capture the turning points in market development, because we simply cannot grasp these events in beforehand. If the risk cannot be eliminated, the question is how the risk can be managed. In other words, it may be impossible to prevent bubbles from occurring – and indeed bursting. Thus, we can only limit the damage. Nevertheless, perhaps AVMs are suitable for it. On the other hand, at the country level there is a clear positive correlation between large variations in house prices and a widespread use of AVMs, which would debunk this argument about any assumed benefits/superiority of AVMs.

The gist of the discourse here is that to connect the sustainability of the built environment with housing market viability is arguably an important topic today (see Jones et al., 2009). However, mainstream market theory offers little help here. Thus the search needs to be directed towards heterodox economic (i.e. institutional, evolutionary, complexity, behavioural and such economic theory perspectives, see e.g. Van den Berg, 2014) or purely practical approaches instead such as the development of AVM for mass appraisal (see Kauko and d'Amato, 2008). This contribution will both discuss aspects of general feasibility and present an empirical application. While environmental-ecological sustainability comprises the most common criteria and social-cultural criteria for sustainability exists too (see e.g. Ahmed, 2011), the emphasis of this study is on economic and – to a more limited extent – social sustainability, as defined as the availability of affordable real estate packages, quality control and product diversity (as well as viability of projects).

The need for sustainable housing consumption

The issue of sustainable consumer choice is to some extent also transferable to a house buying context. However, evidence is yet scarce here. The question of interest is as to w*hat motivates sustainability in housing/real estate consumption.* Possible candidates for answer can be listed as follows:

1   Altruism – to 'do good', so a sense of global responsibility → already in the 1930s Keynes hoped for more of such actors; ostensibly they would be either super rich or of those who have nothing and don't know about better; Kontokosta (2011) suggests that, in general, a small group bears the costs of global carbon reduction.
2   Sheer economics – so here is a profit to be made by 'doing good' (see e.g. articles in the *Journal of Sustainable Real Estate*).

3   Fashion – fits today's trendsetter culture, sign value; this is in between the two previous views.
4   Something else: evidence exists already from China (Hu, 2014): health and comfort rather than sustainability *per se*.

According to evidence from Bangkok (Kongkajaroen et al., 2014), the most important is the option 3. A statistical modelling exercise based on a survey ($N = 220$) of attitudes and intentions of 18–35-year-olds towards purchasing green condominiums in Bangkok applied an analytic separation of five underlying factors: (1) environmental knowledge, (2) environmental concern, (3) perceived benefit, (4) health concerns, and (5) subjective norms (i.e. the influence of others such as peers and family). This study found the fifth factor to be the most important one. The more general conclusion was that certain types of consumers, who can be identified by their more general attitudes and intentions, also are more likely to buy a 'green condo', even if these are more expensive than their generic counterpart. This issue of what exactly motivates the real estate and housing consumer really is an intriguing one with options varying from business and fashion towards altruism and even health related reasons. (Kongkajaroen et al., 2014)

Here it might also be worth considering the difference between two basic groups of demand side housing market actors:

•   Users evaluate what for them is a good building and its vicinity – this is about the quality-of-life (QOL) of the people occupying the dwellings and their daily living environment.[3]
•   Investors evaluate the kind of asset profitability compared to stocks, bonds and so forth.

The need for a sustainable housing market

In the evolution of the functioning of neighbourhoods, the importance of local housing markets is crucial, as shown in the UK by Bramley et al. (2008). On the other hand these authors maintain the importance of house prices having a high responsiveness to income and supply and indirectly being influenced by various planning measures (see Evans and Hartwich, 2005). These factors are partly static and partly time-varying. Furthermore, both macro- and micro-level factors influence urban housing markets. This turbulence can be for better or for worse depending on the economic and institutional starting conditions. Concerns about financial stability implications of developments in the housing market when house price dynamics become disconnected from developments in underlying fundamentals of housing demand and supply are particularly serious in the central eastern

European (CEE) country context, given the strong role of interest rate fluctuations, demographic and labour market movements, development of wages, and establishment of new market and financial institutions (see Égert and Mihaljek, 2008).

Within the framework of a typical Keynesian critique of monetarism, Baker (2008) presents how the financial market is both cause and effect of the housing bubble in the US. He rightly notes that 'the expectation that prices would continue to rise led homebuyers to pay far more for homes than they would have otherwise.' He cites Robert Shiller, who showed that prior to 1995 in the US real house prices were unchanged for 100 years, but by 2002 an approximate 30% overpriced *speculative bubble* completely detached from the *fundamentals of the housing market* was evident. Subsequently, however, the bubble burst – by the end of 2007 real house prices had fallen by over 15% from their peak, which according to Baker amounts to almost 50% of GDP. Baker blames, above all, the appraisers, who had a strong incentive to, instead of valuing an honest appraisal, 'adopt a high-side bias in their appraisals'. Among the reasons he notes the 'complex web of finance that concealed the risk that was building in the financial structure', a large supply of housing being placed for sale and a dampened demand as a result of the stress in financial markets. He finally notes that financial bubbles in principle could be contained (but accuses Alan Greenspan for not doing his job).

Sapir (2008), however, adds the institutional dimension to the criticism in the sense that, if a country is to a sufficient extent detached from the global financial system, it is capable of dampening the effects of crises, and thereby remains unaffected by the credit crunch. The emerging economies on one hand and the continued social-democratic welfare states on the other are such countries. To be able to design better policies than the current ones, we cannot rely on neoclassical economic (NCE) theory, but we need to study more realist and heterodox economic theory (see also Keen, 2009; Söderbaum, 2009; Van den Berg, 2013). Elsewhere, Aalbers (2008) sets up the following argumentation that much resonates with those of Baker and Sapir (cf. MacDonald, 1996):

- *Financialisation* occurs when profit-making occurs in different parts of the economy (including real estate) through financial channels rather than through trade and commodity production.
- Risks in any finance-led regime become risks for all actors involved in any specific industry; hence mortgage loans fuel house prices.
- Mortgage markets are today not only 'a means to an end' but 'an end in itself'.
- Today the real estate markets are more than ever dependent on the financial markets.

- *Vice versa*, the development of financial markets is also dependent on the development of the real estate and housing markets.
- Most homeowners depend on mortgage markets, and this fuels the economy both directly and indirectly.
- This crisis is not limited to the US: many other mortgage markets (e.g. the Netherlands) are increasingly financialised. In fact, a ripple effect to Europe was anticipated.
- The importance of the secondary mortgage market (i.e. investors can buy mortgage portfolios from lenders) has increased over the last ten years.
- The high risk and exploitative character of the mortgage loans.
- Some borrowers should never have received a mortgage loan.
- Not only is 'global' tied to 'local' through of financialisation of the mortgage market, but also the financial and the built environment have become tied together through this mechanism.

As Aalbers shows, the analysis of individuals is used as a basis for credit scoring and risk-based pricing. The key to the crisis is still in understanding the credit scoring of lenders. However, it is not clear why such a criterion could not be based on the product itself – the real estate instead of the personal traits of the lender (see RICS, 2007). In such a case, the role of appraisals is important.

When evaluating inflationary effects against sustainability criteria (see Bramley and Power, 2009), an additional qualification has to be noted. It is true that when house price inflation is higher than the normal index (as measured through incomes and consumption index), the situation is categorised as unsustainable – at least economically – using the definitions of the present study. However, after a lag of a few years the new stock will be adjusted upwards in terms of quantity and quality, which then moves the situation towards sustainability again, albeit at functionally and spatially varying rates.

When examining the extent to which the price development of a local housing market shows a healthy and stable trend, one of the issues to decide is whether the focus of the research ought to be on the location or the built structures. Which one of the two elements then is more important for the investor behaviour? At a more global scale a reasonable aggregate would be 50% for each, but this balance in particular is context dependent. For example, in urban housing market 'hotspots' the location is much more appreciated than the structures, but in derelict brownfield areas as well as for rural property it is the other way around: land being of a minuscule – sometimes even negative (cf. Jones et al., 2009) – value compared to the built structures rather than the site and its location. Traditionally sustainability analysis is biased towards the built structures, but that is for a different reason: because data on the quantitative and qualitative characteristics of the buildings often already exists readily in registers or in any case are relatively simple to collect, whereas corresponding data on the location do not exist and tend to be more difficult to collect. In an attempt to correct this deficit this study therefore looks at the location perhaps more than research hitherto when comparing its value share against that of the building.

When attempting to sort the previous issues, a conceptual umbrella is helpful. What is needed, and what are the consequences of applying any new approach? Here it can be noted that, similarly as the establishment of *institutions* such as the market, the firm and the state has generated economic development through history, at present, new institutions are needed to generate sustainable development. What are then the implications of this for valuations? What is the rationality of valuations? In the environmental valuation/economics literature, Gregory and others in the 1990s argued that one cannot put several value dimensions under one and the same measurement, and as a consequence, multi-criteria decision making (MCDM) gained more credibility since then than the contingent valuation and other one-dimensional monetary valuation methods. However, some problems still persist, especially related to context effects. These are well described in relation to behavioural and experimental economic research (yes-saying, protest bids, etc.) (Vatn, 2005).

Vatn (2005) argues that if only *individual* values are at stake, market valuation is acceptable. However, if we deal with *irreversible* damages to the environment, or even *uncertainty* of such effects, the precautionary principle must be applied: that is moral considerations are valued higher than cost-benefit analysis and no scientific evidence should be required in order to apply the best measures available. This brings us to the normative role of valuation. The difficulty is that if multi-criteria decision making (MCDM) is preferred over cost-benefit analysis a lot of complexities ought to be taken into account, which in practice is rather difficult.

Conceptualisation of the suggested improvement

The comments in the previous section suggest that the way we understand and explain property value might be undergoing a paradigm change. However, several issues mark this (plausible) change, and to dwell on all of them is beyond the scope of this contribution. The way 'value' and 'market' are understood from an epistemological or ontological point of view must nevertheless be of concern here, given our ambition of improving the methodological basis for valuation. In the old paradigm the separation of the various dimensions (economic, physical, social, cultural, etc.) was acceptable, because being able to use equilibrium economic models such as the hedonic model, which would not have been possible otherwise, was considered the goal. In the new paradigm the methodology might be pragmatic instead of formal, but that is not the point; the point is to avoid a totally artificial separation of dimensions as is the case with the models and methods of the old paradigm. In doing so, it puts a further emphasis on the behavioural and institutional elements as well as on a communicative purpose in disseminating the results (rather than the instrumental purpose). In other words, the motive is holism rather than reductionism (cf. Van den Berg, 2014). In doing so, the view propagated here purports that in the new paradigm an approach to valuation methodology ought to comprise the following five main elements: (1) sufficient detail in the analysis, (2) the goal in the long-term situation, (3) the recognition that a long-term diversification and dynamics moves us towards sustainability, (4) the use for relevant market sustainability data, and (5) the flexibility to allow for an uneven change towards sustainable markets.

Detailed analysis

If the aim of the study is detailed analysis of the market for property products and locations, for example, in the mould of the studies of sustainable housing consumption referred to earlier, it is always important to incorporate *a diversified process/dynamics view* for the demand side: that is diversified and changing preferences of buyers and renters as well as intermediaries need to be recognised. Otherwise we manage only to capture an average type of market behaviour at best, and a miscalculation at worst.

The long-term situation

If the topic of analysis is the long-term situation it is similarly important to incorporate a diversified process view for the supply side analysis (sellers, investors, builders, developers, planners, etc.); otherwise, we fail to see the effect of factors such as additions/reductions in the stock, outcomes of planning, demographic change, increased/decreased mortgage availability and new consumption patterns. For shorter terms, however, merely a *diversified static* view (i.e. a collage of snapshots at a given point in time) suffices.

Long-term diversification and dynamics is sustainable

As long as the market analysis concerns the long-term market situation, economic sustainability becomes a natural issue. This is not an attempt to attack the economists' defensive mantra of 'in the long run we are all dead', but more constructively, to realise the added value of looking ahead – not necessarily more than to the next generation (or two) of housing consumers. By the time these people enter the markets most of the housing stock and housing environments will be the same as we see today. It is to observe that, due to the context dependent normative element – that is to say, regulations vary and change – here it is particularly important to look at diversification and dynamics, even beyond the current range of standard housing market products.

Data on market sustainability

The previous arguments all suggest that to analyse the market and sustainability poses huge challenges for the production and evaluation of datasets. Even more so if non-economic dimensions, that is to say ecological, environmental, social or cultural values and benefits that we can identify but not necessarily quantify, are included. Not only data on prices, rents, yields and financial-economic variables are needed, but also on physical, aesthetic, health and other characteristics of the stock and environment, even though we in many cases have to accept the subjective and hypothetical nature of this data. Optimistically market actors are expected to generate such data and to make it available for analysts in times to come. A good start would be to satisfy the need for quality control and availability of house price datasets (Francke, 2010).

The change towards sustainable markets is uneven

Due to the previous point: the need to collect relevant data, it can now be argued that the analysis needs to begin from context and then select the methods based on the available data, rather than the other way around: screen the data based on assumptions as is the case with NCE oriented analysis. In other words, one should proceed bottom up and employ the richness of empirical material instead of trying to fit all circumstances into the same model or methodology of appraisal/valuation. This is even more the case for sustainability evaluation, which, as already argued, necessitates adding dimensions and taking the longer term perspective compared with market analysis. Or, we cannot treat industrialised and developing countries the same way; neither can we do it with urban and rural areas, or with market and public housing; and so forth. This logic is illustrated with an example Figure 2.1, concerning the issue of affordability. According to a general proposition by Wallace et al. (2009) the sustainability of homeownership is threatened by high housing affordability ratios. The specific problem here is that the asking prices for row houses in the western and central regions of the

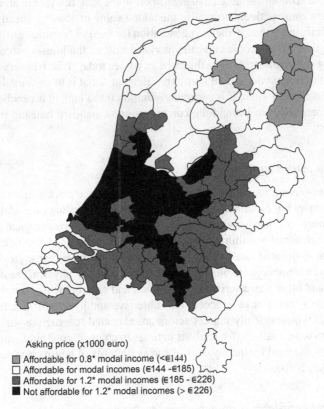

Asking price (x1000 euro)

■ Affordable for 0.8* modal income (<€144)
□ Affordable for modal incomes (€144 -€185)
■ Affordable for 1.2* modal incomes (€185 - €226)
■ Not affordable for 1.2* modal incomes (> €226)

Figure 2.1 The affordability of standard row houses for three income levels (modal income, 20% below and 20% above modal income) in the last quarter of 2006.

Source: Op't Veld et al., 2008

Netherlands are too high compared to modal income in year 2006; if we take a long-term view the evaluation is that the housing market is economically and socially unsustainable for that segment and part of the country – but not in the country as a whole.[4] Similar concerns about the prices becoming unaffordable in certain regions have recently been aired by others in the field, for instance, Cheshire (2005) about London in relation to the rest of the UK, and Lux et al. (2008) on Prague in relation to the rest of the Czech Republic.

Current AVM practice

Given the often doubted relevance of AVMs for valuation and market analysis, the next question is whether the occurrence of recession since 2008 has had an impact on research and development (R&D) activity on AVM. To find out a good starting point is the recent literature survey by Downie and Robson (2007). From their comprehensive literature review on automated valuation models Downie and Robson find that traditional valuation approaches are being increasingly replaced by the cheaper and quicker AVMs, and that this development goes together with the need to control loan decisions. They identify, however, potential problems due to inaccuracy and the fact that AVMs have not yet been fully tested in a housing market slump, although some evidence from the US indicates that circumstances of market downturn do not cause invalidity. Furthermore, in a more established market best practice guidance is being developed about *when* and *how* AVMs are to be used. Finally, Downie and Robson argue that human valuers will not become obsolete even if many valuations are carried out electronically, as the US experience shows. They conclude that standardisation of AVM procedures and features is the crucial issue that determines the direction of R&D in this respect, and that there is potential for proving the benefit of AVMs for stakeholders in less established markets.

To get some insight into the discourses being recreated within this realm, the following brief review of currently available American and British commercial AVM options is informative. A variety of notes, columns and adds written 'with post-bubble hindsight', that is since the summer 2008, in the recently established e-newsletter *AVM News* (2008a, b, c, 2009) are used as a source. While this documentation is largely opinion based, the first fact to note is that the Obama Administration rather immediately included appraisal and valuation components in its guidelines for the Home Affordable Refinance Program, and a number of federal agencies have jointly issued the following comment:

> Volatility within certain real estate markets and associated credit risk underscore the importance of independent and reliable collateral valuations.

Elsewhere, it is clarified that lenders are apparently more interested in the capability of the potential borrower to repay the loan than on the current market value of the property. The AVM development framework furthermore seems to be a local rather than national issue according to some experts; this is after all understandable given that housing markets are granular and localised. Anthony Garritano, Editor of *Mortgage Technology* magazine, points out that it is ultimately the investor who determines the value, and that technology cannot work alone without the control of lenders and servicers because the risk in relinquishing the collateral values to a model or third party is too high. He states that, while there is a role for predictive models, the role for solutions that provide transparency and data management capabilities is even greater, and that 'the real need is for accurate analytically based valuations combined with solutions that allow users to understand the different value opinions'. It is furthermore noted that several different tools or methods should be used for valuing the same property or to support a particular lending activity, and that an institution should not select simply the one that provides the highest value, but instead establish criteria where property type, location and the nature of transaction all are taken into consideration. Finally, it is speculated that the future of valuation lies in a blend of 'traditional human evaluations' and 'automated products'.

In the US several technology and database companies have taken the AVM technology on board. *Zaio* provides an instant predicted market value as well as an estimated value based on the assessed value of the property using digital photographs of the property. This can be used as a tool to check property values. They maintain a database of 140 million properties. Their goal is photographing every home in major cities and metropolitan areas. The recent alliance between *FNC* and *IntelliReal* gives a sophisticated option for AVMs as this product utilises multiple listings data from 77 million property records nationwide. *Visre* is a company active in imaging for country government tax reassessment projects in Louisiana and Georgia. Their vehicle mounted camera arrays that enable imaging 4,000 parcels a day offer a service for reducing the risk levels of the valuation. *Realtor. com* offers yet another tool that utilises multiple listing services. *Lender Processing Services* (*LPS*) is a particularly interesting option as they have announced a launch of a system for value forecasting at the neighbourhood level, including delinquency and inventory trends for the immediate neighbourhood surrounding a specified property. *LPS* thereby has a solution for understanding not only the property's current value but also its likely direction in the near future, given that individual property values are substantially influenced by neighbourhood characteristics. Other appraisal systems worth mentioning here are 'the three leading home valuation sites': *Zillow.com*, *Listingbook.com*, and *Trulia.com*; the San Diego-based *MDA DataQuick*; *Boxwood Means*; *Patriot*; *Integrated Asset Services* (*IAS*); and – this seems somewhat obscure even within the esoteric AVM community – a cooperation between the firm *Smithfield & Wainwright* and an anonymous 'Florida State University alumnus'.

AVM has become increasingly common also in the UK residential mortgage markets due to a rise in demand for mortgage products for which lenders require a more time-efficient and cost-efficient valuation approach than the physical inspection undertaken by surveyors which hitherto has been the standard practice (cf. Op't Veld et al., 2008, on the corresponding situation in the Netherlands). *UK Valuation*, the UK's pioneering provider of AVMs, reports that their AVM solution is performing robustly despite the continuing decline in UK house prices: their largest observed undervaluation of test portfolios during a more than two and a half year period was just 0.5%; their parent company *First American* recoded an overvaluation of 1.7% during a half year period of falling house prices. Moreover, their market-leading AVM solution *Mortgage Brain* has launched a systems' integration where users get instant access to its AVM using innovative data modelling techniques to enable provision of residential property valuations in service characterised as highly accurate, fast, efficient, complete, simple, and easy to use. *Moody's* modelling approach also quantifies the risk associated with the use of different AVMs in Residential Mortgage-Backed Securities (RMBS) transactions. *Hometrack*, which claims to have 95% of the AVM market in the UK, is claiming to reach acceptable levels of accuracy irrespective of whether the market is rising or falling.

Based on this documentation of the practitioners' state-of-the-art – sometimes referred to as 'the dark side' by academic researchers – there seems no end in sight in the drive to establish innovative solutions for AVM, and if anything, the crisis has only accelerated this development.[5] Such aspirations align with the aim of this book, designing an empirical approach to determine the degree of urban housing market sustainability, that is the extent to which prices, rents and turnovers display self-supporting qualities and retain them in the longer term (cf. Jones and Watkins, 1996). The next task is therefore to look at the issues at stake from a 'more scientific' point of view.

Designing an empirical modelling method

A similar ingenuity as the practitioners have is yet to be seen on the academic side. Within the residential sector two different broad scientific approaches or perspectives to valuation and market analysis exist: one, hedonic, which has been extended in recent years; two, the alternative or non-market perspective which is yet in a testing stage. The approach propagated in this book falls in between those two but is closer to the latter. Drawing on prior contributions by the same author (see Kauko, 2001, 2004a, b, 2008a, b, 2009c; d'Amato and Kauko, 2008), the point here is that context often is an important precondition to take into account when selecting/designing valuation methodology in relation to specific changes in the market environment. Furthermore, we need to examine two different groups of procedures: group 1 concerns *direct* comparison of known cases (i.e. the standard comparable sales approach); group 2 concerns hedonic regression or other type of *indirect* valuation methods and techniques that, based on specified and estimated models, inform about coefficients and adjustments for the former group. Note that the latter group is rather reductionist and thereby rather problematic for the argument insofar as we wish to take distance from a hedonic regression approach.

As worldwide developments have shown, issues related to the economic sustainability of local housing markets cannot be tackled by mainstream housing economists. Instead the agenda needs to be reset. In doing so it is argued here that we might look at heterodox economists on the one hand and practitioners on the other to find helpful ideas, concepts and research attitudes. At any rate, to connect discussions on housing market viability with discussions on built environment sustainability (see Bramley and Power, 2009).

This chapter has documented a number of key points that require attention when designing a methodology for evaluating local market sustainability. Given the backdrop of the financial crisis worldwide and housing market downturn in most western countries, regions and cities, the list of practical AVM applications launched for valuation is astonishing. Throughout the community of leading commercial AVM developers various innovative solutions are being tested 'as we speak' – it is all about going forward in this problem field regardless of any doom-and-gloom scenarios. This really is an unexpected finding. The present study has attempted to catch some of this attitude and take distance from the current conservatism of the academia in this research area. That the frontier of valuation research is moving forward so rapidly is an embarrassing finding for us academic housing market and real estate analysts. This brings to mind the conclusion by Söderbaum (2009), who argues that we academics cannot escape the blame for causing the economic downturn, because one of the roles of the science and university is to inform actors and governments about the right decisions to make. In particular, NCE has contributed to legitimate neo-liberal policies and, as these policies have failed, so has the teachings of the economics departments. While the hardest criticism can be levied on general economists, to a lesser extent the same applies within the applied fields of real estate and housing economists.

However, we need to be constructive: if market sustainability is a goal, what kind of methods and models should we then aim at? The desired level of empirical model performance depends on the context, but also on the preferred balance in trade-off between conceptual soundness and accuracy. This is why the sufficiently accurate linear hedonic regression modelling of the housing market has proven reasonably successful. On the other hand, alternative approaches allow researchers to capture the complex nature of the housing market relationships. In this body of literature the *methods in relation to data* are categorised onto the following typology:

- Statistics in a parametric distributional sense, which then validates the hedonic model;

- Other statistics and related approaches (e.g. the SOM);
- Other data, that is judgemental and possibly more qualitative approaches:

  - When we for one reason or another need different information, and therefore need to collect data for new variables;
  - When we don't have information even of the standard variables (Kauko and d'Amato, 2008).

In this study quasi-dynamic 'models' are created using Trondheim and Amsterdam housing market data and a time-window approach to using the SOM, following Carlson (1998). Due to its pragmatic and non-linear nature the SOM time-windows approach to quasi-dynamic market modelling seems promising in this respect. The findings also have policy implications: using innovative empirical housing market modelling as an analytic tool enables planning for sustainable land use and governance on an urban and regional scale (cf. Bramley and Power, 2009). That said, the potential problem here lies in academic trustworthiness; this approach might seem too different from an economic mainstream approach (see Lorenz et al., 2008).

The economic sustainability thesis within a house price analysis context

Among several vaguely defined, non-standard and competing definitions for economic sustainability, the study applies a rather pragmatic stance – the *value stability* concept. This is built in congruence with the research design and methodological approach outlined in Chapter 1 of this book:

- As grossly substandard level of housing is unacceptable for health and safety reasons, the quality (largely a subjective indicator though) ought to develop in the same direction and with the same pace as the price level; this is about the site and building specific attributes as well as the characteristics of the surrounding environment, neighbourhood and the city as a whole.
- As it is not sufficient with high quality unless people can afford to buy (or rent) it, the affordability (often approximated as net income) of the dwelling also ought to develop in the same direction and with the same pace as the price level.
- Even if the quality and affordability criteria are fulfilled, it is not sufficient for value stability (and hence economic sustainability) unless there is a wide enough range (i.e. product variety generated for most apt selections to be made) of different quality and affordability levels on the market. This is because the drivers of sustainability: production technology, community governance and consumption fashions all tend

to change quickly and then it is vital not to have neglected any specific property/housing package even if it may seem marginal at some stage. Or put it differently, if a potential market trend setter or other innovation in terms of quality or affordability is not recognised this will have harmful impacts for the evolution of the property portfolio in terms of its value stability.

At this stage it can be concluded that socio-economic metrics, including market value estimated for 'green' features, would be helpful for market analysis and valuation science. Furthermore, this opens an opportunity for the development of AVMs to capitalise on.

Notes

1 This is to paraphrase one of the panellists at the EXPO REAL 2008 in Munich.
2 It is for example argued that sustainability can be evaluated only to the extent two of the three criteria overlap (Silhankova and Pondelicek, 2010).
3 For example, Rauterkus and Miller (2011) promote mixed use developments as these are walkable in relation to neighbourhood amenities and also bring land value premiums in such neighbourhoods.
4 Elsewhere, Francke (2010) applies error correction modelling to evaluate whether the Dutch (nationwide) housing market and house prices are overvalued and finds out that this is not the case. In year 2010 the price development is only 1%, which is about the same as inflation.
5 The same story continues in later editions of this e-journal. For example, in *AVM News* (2011) the UK based property portal *Zoopla* has launched a methodology for tracking home values in a market situation where they have fallen 11% during three quarters and 18% from the peak. Moreover, in many of these marketing comments the argument is about the developing of the 'next generation models' that integrate an element of third-party inspection into the analytics. This resonates with Kauko (2004c). And most recently Peter Stimson, who is Managing Director at Landmark Information Group, notes the value of what he calls 'green' AVMs (*AVM News*, 2014, p. 9).

3 Assessment criteria for sustainable local price development

Either proactively or reactively, approximately, since the turn of the Millennium, in many urban areas real estate investments are considered in a more sustainable framework than before. And this is indeed important, given that real estate comprises a lion's share of world wealth – at least before the credit crunch of year 2008 more than 50% of total assets (in the US, 75%) was tied in real estate. Because of the recognition of this relative importance, the real estate markets have been analysed in various European countries since the 1950s through the 1960s, from both points of view: academia and practice. The former analyses have pertained to hedonic price/market modelling and other kinds of scientific analyses, whereas the latter often has been subject to a normative approach. Still real estate research is a relatively neglected, trivialised and under-theorised problem area pushed at the margins of economic and technical disciplines. With the increasing global significance of the sustainable development agenda this neglect needs to be corrected, given that the real estate field deals with three dimensions of sustainability: the economic, the social and (physical) environmental one (see Bramley and Power, 2009).[1]

While sustainability was earlier an issue for academics and environmentalists, today the public sector has incorporated sustainability goals. Now that public-private partnership (PPP) is an issue in property development it also has become an issue for private sector and property markets. It is recognised hat joint ventures can open up niche markets, which then creates an opportunity for property developers (Bryson and Lombardi, 2009). The big issue is of course: how to achieve this sustainability? It is to observe that a sustainable building in principle can be attractive for both owner and tenant. However, the literature on urban real estate sustainability has not yet reached an agreement as to whether sustainability always would be a worthwhile business strategy. This depends on the extent to which the extra costs[2] can be covered in the long run from various benefits such as improved image of the firm, the security of being able to provide apt and attractive facilities for the end user – tenants as well as work force. On the other hand, corporate social responsibility can be used as a pure business strategy without much altruism, as experimental economics and games have shown (Bénabou and Tirole, 2009). It should also be kept in mind that the issues are not only restricted to energy efficiency and other green aspects, but also economic and social issues need to be rethought within the urban real estate sustainability agenda.

In Chapter 1 the socio-economic side of sustainable real estate was noted as a less covered topic than the 'green' real estate issues. Whereas metrics for the latter context have existed for a long time already, similar measures for the former is yet to be developed. On the other hand, price is a handy indicator as it includes plenty of information about the market and thereby already tells us something about the sustainability of the particular case. On the other hand, by the same token it can be argued that price alone rarely is a sufficient indicator and other measures to relate price to are required. Preferably all this is set in a spatial setting, although there is no requirement for an extremely sophisticated technique unless it brings added value to the task at hand. And bringing added value to real estate analysis is the ultimate goal here.

When linking *the greening of the urban property industry* with other economic sectors, one of the important conditions for achieving a greener economy is the education of consumers and construction industry professionals. This education can change the behaviour of housing consumers and energy saving of the households will lead to greener housing and residential areas (cf. Wilkinson et al., 2013). On the other hand, education of the producers and professions is vital too. The challenge of the construction industry is to come up with innovations, not least for retrofitting the buildings, which in turn generates greener housing and residential areas too. This in turn will generate greener jobs, which will accelerate the process of greening the economy. This can then be linked further to the demand side processes insofar as the consumers, through their other environmentally and socially responsible behaviour, lifestyles and daily routines, contribute to sustainable consumption, and thereby to the greening of the economy further (see Spaargaren, 2003).

On the other hand, the history of the global sustainable development agenda is notably short. Since the oil crisis and the predictions of the Club of Rome in mid-1970s the overconfident forecasts of unlimited growth began to change. A cut down in energy consumption necessitated a more proactive approach to the use of resources. Yet on a large scale, a major turn in consumer attitudes and political rhetoric and, perhaps more important, a shift in the adapting of new technology, began as late as in the early 1990s. Since then sustainability has been promoted as a key overall criterion for steering physical development. In particular, the international sustainability development agreements include poverty eradication as a vital element (Bramley and Power, 2009). Unfortunately, the meaning of this concept is still obscured by the various discourses – some of them more and others less scientific.[3]

While accurate definitions as well as hard evidence of sustainability within the real estate field are still missing, there is reason to believe that in many urban areas real estate investments are now considered in a more sustainable development framework. We may be witnessing just the latest fashion or then a fundamental paradigm change but, either way, the attention of property investors and government regulators alike is being shifted from mere economic efficiency goals towards long-term and multidimensional sustainability goals. As in many cases a new investment cycle in the built environment is about to commence, the question is whether the investment could or should be economically, physically and socially sustainable. Then the next question arises about the possibility, feasibility and necessity for either new development or refurbishment of the building stock. The extent to which such issues are – or even can be – on the agenda depends on the character of the area within the city (or city-region), the city (or city-region) itself and the institutional setting where investment takes place. Furthermore, it is about the quality-of-life (QOL) of the people occupying the dwellings and their daily living environment (cf. Fahy and Ó Cinnéide, 2008).

In sustainability issues concerning land use and real estate, the government has a twofold influence: partly traditional, through tangible infrastructure provision (public cost-benefit, measured using hedonic impact analysis) and partly more modern, through cultural consumption and image creation for city competition (see e.g. Glaeser, 2011). From both roles it is possible to build a link to a variety of aspects related to sustainable urban real estate – residential and commercial alike. The issue at stake is that of agreeing on an exact definition. Is sustainable development in an urban context environmental-ecological only, or can it also – and alternatively – be social-cultural, or even economic-financial? In this contribution the economic criteria is selected for scrutiny. This is because of the apparent need to work on the relative obscurity of this concept and subsequently filling the void of such research. (More on this later.)

The next two sections of this largely theoretical chapter discuss possibilities for creating a sustainable local property market. The first section deals with real estate strategy generally, whereas the second section focuses on the special role played by the location. The third section outlines the ontological and epistemological basis of a methodology for empirical sustainability assessment.

In search for a sustainable urban real estate market strategy

Market context vs. sustainable development

As an investment class, urban real estate offers some promising prospects for conforming to sustainable development. To date, however, the evidence is scarce and speculative at best. Real estate economists nonetheless see enormous potential in sustainability. According to estimations by Lützkendorf and Lorenz (2007) a 10% portfolio switch in property assets to socially responsible investment (SRI), which recently is taken into use mainly in the US and Australia, would generate a 50% market capitalisation of the *FTSE EPRA/NAREIT global listed real estate index*.

When investigating a related concept, corporate social responsibility (CSR), Cajias and Bienert (2011) emphasise the importance of the participation of real estate companies in international development and sustainable policies. In their econometric study across European listed real estate companies they find evidence in favour of two promising tendencies. The first is that as incentives to disclose information of CSR lead to improved financial transparency, CSR can be seen as an innovation that contributes to sustainability. They observe that the studied firms on average increased their CSR activity between the years 2007 and 2008. The second finding in the study by Cajias and Bienert is that CSR lowers risk and uncertainty (cf. Bénabou and Tirole, 2009; Colantonio and Dixon, 2011).[4]

Risk obviously is a major issue here and an emerging research tradition is already attempting to unveil this aspect. Jackson (2009) evaluates the riskiness of LEED and Energy Star development and concludes that the latter category of green projects provides both greater returns and lower risk and is thereby to be considered the more attractive option of the two (cf. Mehdizadeh et al., 2013). Galuppo and Tu (2010) in turn carry out an online survey of the perceptions of lenders, equity investors and developers regarding green buildings, and find that the growth of green development is being hindered by the lack of consumer awareness of benefits of building green, along with lack of incentives.

Elsewhere, housing and planning research address similar concerns. The key is *containment of land use* in relation to *the preferences of consumers* (Reese and Sands, 2007). It is noted, among other problems, that contrary to popular belief also the upper market housing demand may be unsustainable: increased suburban detached homes and car ownership and frequent trips is not sustainable from a more holistic point of view, even though efficiency of resource allocation may be met for certain segments (cf. Bontje, 2004; Pareja Eastaway and Støa, 2004; Stenberg, 2008).

While it is reasonable to assume that even government structures will eventually adapt to sustainability criteria, the immediate aims are likely to be more of 'bottom-up' than 'top-down' character.[5] Arguably, the most sensible approach to assess suitability within urban real estate market context is to look at how organic change, as opposed to government induced change, is possible by convincing investors, developers and homebuyers about the needs to engage in sustainable real estate strategies.[6] This is not to deny a role for government. This role is, however, largely in stimulating – rather than regulating or direct provision of – conversions, refurbishments and new developments into sustainable modes. The pivotal issue is in other words about educating the mass of real estate actors to voluntarily apply sustainability thinking. The strategy of sustainable developers is to reap normal profits from the real estate development, and then feed the remaining profits onto 'use value' considerations such as improved infrastructure, availability of more green areas and providing a certain percentage of affordable housing (Bryson and Lombardi, 2009). However, this principle is debated, and it is to note that the discourse is rich and evolving.[7]

The view currently propagated by some experts is that policy making must be about 'smart regulations'. It is not just enough if we have enlightened developers such as *Urban Splash*, *ISIS* and *IGLOO*, as we need patterns of governance too – and policy makers can be smart developers. Thus the rules and regulations need to change in order to achieve a sustainable real estate market in the long term. One is, however, left wondering: is it really only about regulation and not about behaviour? Arguably the juxtaposition of 'long-term view' and 'regulating' is a contradiction; is a certain regulation which is apt today still going to be apt in twenty years' time? It is unlikely, given that regulations and policies tend to be more conservative than the behaviour of individuals (Ball, 2006); the path-dependency has a proven omnipresent influence (see Martin and Sunley, 2006).

The complexity and contradictions of the discourse is obviously a challenge here. While 'sustainable real estate' is a well-recognised topic, the views on sustainability in this field are much divided between those who believe in the responsible and enlightened behaviour of developers and investors as well as consumers together with institutions that stimulate such behaviour, and those who only believe in regulation of technology and consumption. We could say that the former view is 'the elitist view' (or naïve view, for those who don't share it) and the latter is 'the cynical view'. While the latter view is inconsistent in trying to address long-term problems with policy/regulative design, they do have a point in the sense that the very promising 'elitist' thoughts aired primarily in German speaking and possibly Benelux, Nordic and some Anglo-American countries might not be easily transplanted to other contexts. Even the UK experience shows that practice does not easily follow state-of-the-art research. There could be room for convergence for these two views, however.

The 'elitists' believe that we need 'supportive rules of the game', so that market players can achieve better prospects of reaping the benefits of sustainable design and buildings. This is much about taxation. For instance, if one receives a real tax rebate for one's property having a positive energy performance certificate, the whole discussion about mandatory vs. voluntary certification is ended, because everyone will want to have a positive certificate as the 'polluter pays principle' is put into practice (e.g. González and Kern, 2007). Moreover, some large property investors and owners (in fact, also the *European Property Federation*) indeed see the financial benefits of energy efficient or sustainable buildings.[8]

Related to this need for clearer guidelines (or sanctions), Sayce and colleagues (2007) underscore the importance of sustainability incentives such as taxes as well as the development of a sustainability metrics. There is widespread support among stakeholders for fiscal measures that might incentivise a movement towards more sustainable property investment and management in the UK. Sayce and colleagues point out that it is difficult to create a business case for 'sustainable property' as the would-be benefit would have to come from cost reduction, more profit or higher value, or risk reduction; currently only the last benefit really exists. Furthermore, without appropriate metrics, the business case lacks transparency for the investors; an index would generate credibility, which subsequently would lead into market transformation. Fortunately, their own questionnaire survey evidence points to a slowly changing behaviour in this respect. An interesting further issue brought up by Sayce and colleagues (and which already was mentioned in the first chapter) is that the still vague but evolving discourse of social and economic sustainability shows more resonance with investors than that of the older type of environmental-energy sustainability. This gives us confidence to delve into social and economic matters at the expense of the well-trodden paths of ecologic and green arguments.

On the other hand, Wagner and colleagues (2007) reach even further in this realm. They show that 'green buildings' (specifically the building performance of low energy offices) do not contradict social sustainability that results of increased occupant satisfaction. In fact, this does not even contradict the economic sustainability that indirectly results from increases in worker productivity either. Here empiry of sixteen buildings in Germany suggests energy efficiency (thus the 'green' dimension) and occupier satisfaction (thus social and economic dimensions) work in the same direction! Similar findings from other countries and real estate sectors would be welcome to generalise these findings.

When delving deeper into the debate between the elitists and cynics, we inevitably also connect with another debate concerning hi-tech and ICT oriented new development projects. The *Smart City* approach to sustainable development is also increasingly considered in several countries including all EU member states. This approach is a 'mixed bag' of strategies that go beyond technical improvements, even though putting fibres so as to increase the internet accessibility of the whole community is a key part of it. Cases of Zurich and Edmonton are good to mention here, as in both cities such applications are already in use (Zurich: an application that synchronises ambulance services and another application that reports potholes or broken street lights to the council). It is about GIS mediated real time data transfer between user and provider; using sensors, satellite technology and mobile devices this technology will be increasingly common in the future (see Perspectives, 2013).

Proponents of this approach argue for a real estate development principle where cutting edge technology is utilised to the maximum extent for modernisation of the urban realm. Those who are staunch advocates of this approach see a promise in the way real estate developments are directed by real time data that increasingly become better available, and this, in turn, would foster democratic and inclusive participation. Furthermore, in this discourse a connection is made between agglomeration benefits and urban development principles based on human capital accumulation and technological change (Glaeser and Resseger, 2009). Well-known examples include the *New Songdo City* in South Korea, *Masdar City* in Abu Dhabi and *PlanIT Valley* near Paredes in Portugal (Greenfield, 2013). These are all mega-projects on greenfield land, none of which is completed at the time of this writing.[9]

Wheeler and Beatley (2014) discuss the first two of the abovementioned *Smart City* projects. *Songdo* is characterised as follows:

- The lot: 1,500 acres of marshland and bay fill;
- The first phase (2009) comprises *c.* 100 buildings;
- The price tag of the project: 35 billion USD;
- Developed by Gale Int. (New York based) and Posco E&C (Korean); Cisco cares for the infrastructure;
- Sustainability claim: LEED certified;
- Criticised for loss of wetland for birds.

Masdar, in turn, is characterised as follows:

- The lot: nearly one square mile in the desert, *c.* 20 km from Abu Dhabi;
- Integration of modern technology and ancient architectural practices from desert environments;
- The price tag of the project: 20 billion USD;
- Developed by a government-owned development company;

- Sustainability claim: 80% of water is recycled, photovoltaic panels and so forth;
- Criticised for being built by 'Petro-Dollars'.

However, not everyone are convinced that this is a positive development overall, for example, Greenfield (2013) is diametrically opposed to the basic idea of the Smart City approach, arguing that it is nothing but yet another unjust, unethical and an unsustainable business model, where real estate development is married with hi-tech solutions that are much based on surveillance data on ordinary citizens – by definition something that is anti-democratic and elitist.

Related to this, another debate with much similar content is that between two different governance ideals (good government being one of the cornerstones of urban sustainability, as discussed elsewhere in this volume). Sager (2009, 2010) separates two modern planning paradigms or, more precisely, two modes of governance, as follows:

- Communicative Planning Theory (CPT):

  - multidimensional values
  - about citizen satisfaction and quality maximisation
  - public inclusion and deliberative democracy is important
  - and responsiveness to other parties even more so
  - The weakness of CPT is that it is time consuming.

- New Public Management (NPM):

  - neo-liberal planning theory
  - monetary values (often cost minimisation)
  - about market rationality of the public sector too
  - managerial accountability based on performance is important, which requires specific quality measurement
  - Questionable whether NPM is still democratic.

In the same vein as previously described, this debate NPM vs. CPT is simplified into two competing ideals – one is seen as business and technocratic and the other as community oriented and democratic.

According to Sager (2009, 2010) the weakness of CPT is its accountability of responsibilities. However, governance transparency is an undisputable benefit that may well counterbalance any such weakness. The theory thus suggests collaboration between key actors: investors, contractors and regulators – and ideally also residents, as a valid criterion for judging urban property development to be sustainable (although in reality the collaboration might be limited to a NPM framework where simply a well-managed PPP structure is considered sufficient, see Sager 2009, 2010). Sager (2009, 2010) much upholds the dichotomy of the two ideals: NPM, while accountable and efficient, he sees as having too much managerial control, whereas the CPT, in turn, he declares transparent and democratic despite neglecting the professional autonomy considerations. Sager nevertheless observes that these two types of governance/planning can be combined into one and the same NPM + CPT model.

Obviously, these kinds of debates between sustainability based approaches are more or less consistent with the borderlines of more general debates within social science. The most general one is that between neo-liberal and Marxist inspired understandings of urban development. It can be argued that, while a zero sum game with inbuilt destructive automata of capitalism in the sense postulated by the neo-Marxists clearly is an invalid conclusion, the sustainability in this context depends on the education of consumers and professionals as well as on the implementation of smart regulations and economic incentives. Obviously, the issue is to validly criticise neo-liberal government policies. The proposition here is to replace this antagonism between neo-Marxists and neo-liberalists with a genuinely sustainable development view in relation to institutional, evolutionary, ecologic and complexity economics and human behaviour (see Foxon et al., 2012). Thus the overall stance taken here is that, when demonstrating the local opportunities for economic gains it can be argued that, a critique of neo-liberal politics is necessary but does not equal following a neo-Marxist approach.

Market value vs. sustainable development

The issue about greening the real estate markets has received attention recently from a largely neoclassically trained community of real estate economists. Fuerst and McAllister (2011) produce quantitative evidence about the impact of green certification on commercial real estate prices. Using a hedonic approach and a comprehensive panel dataset of observed prices (9,806 sales during 1999–2008) and rents (18,509 asking rents of Q4 in 2008) from US cities, they conclude that eco-certified buildings (CoStar database: LEED and Energy Star certified buildings) have both a rental and sale premium. The rental premia for certified building were reported to be 4–5%, whereas the price premia were found to be as high as 25–26%. However, this study arguably suffers from two major weaknesses: first, it applies a NCE approach which only takes into account sustainability in a narrow, traditional sense;[10] second, it applies nationwide data in one and the same hedonic model which is inconsistent with the utility maximising idea of the microeconomic framework that hedonics fits into.

These findings are worth comparing with other similar evaluations. Using data of certified commercial green buildings in the US – Energy Star and LEED (Leadership in Energy and Environmental Design) – Eichholtz et al. (2009, 2010) find lots of evidence to prove that sustainable buildings really command a price premium (per square foot, for Energy Star) of the following magnitudes: 3% for contract rents, 7% for effective rents and 16% for selling prices. These findings indicate that the positive price effect goes beyond 'mere intangible labelling effects' (as Eichholtz et al. put it).

Similar results have been obtained from Singapore, where Deng and colleagues (2010) find evidence of 'substantial' economic returns from a Green Mark residential building introduced in 2005 by the country's Building and Construction Authority. They use a dataset of over 36,000 transactions and a two-stage hedonic model. In an updated study using the same database and geo-statistical hedonic modelling, Yu and Tu (2011) confirm this positive price effect, and in addition speculate about the origins of it. They find out that the premium might be due to higher building costs rather than about green premium in itself. This is because they find that the ostensible premium is higher for new buildings than for second hand markets.

In Hungary Farkas and colleagues (2004) reached similar conclusions as the abovementioned regression studies. Their results are based on a survey of *c.* 12,000 owners of homes in 315 settlements of Hungary, where the given property characteristics were recorded. In this survey the owners were asked to assess the value of their house. The impact of energy related property attributes on per m^2 price varies between +9% for boiler heating to −14% for flats with room heating with coal, wood or electricity. The explanation lies in the future affordability. When looking at the housing market, there is sustainability premium for modern flats with lower energy costs. The energy costs are especially high for panel buildings. These flats are as a consequence difficult to sell as the affordability of their maintenance is weak.

In Italy Massimo (2011) finds premiums of as much as 27–30% in favour of a green building in comparison to a building that complies with the minimum regulations (brown building). The sample size is small, however. Elsewhere, Thiet (2011) points out that, whereas no costs are added and the value is ensured by such features, there are no price or rental premiums at the moment due to the market slump. However, as this is a specific situation he believes that there will be premiums in the future. In a similar vein, Schumann (2010) finds that sustainability in general increases the investment value in Germany but that the major uncertainty involved makes her sceptical about the viability of sustainability investments in Germany.

Going back to Singapore – where a particularly well-accumulating evidence base is found – Addae-Dapaah and colleagues (2009) carry out a survey of commercial building owners' perceptions ($N = 400$), using a *Likert* scale and *principle component analysis*. They find out that direct benefits for the business economic performance rather than environmental concerns affect the willingness to occupy 'green' buildings. Among such benefits they list environmental factors, productivity gains, improved internal conditions and cost savings (in declining order). The respondents also express some doubts, notably, that perceived benefits such as improved workers' productivity and user satisfaction are subject to a higher degree of uncertainty than if they were measured from actual outcome data; and that even the environmental benefits do not induce a greening of the behaviour of those who consume the commercial buildings. As a remedy these authors propose the possibility of governments to offer tax and/or financial incentives that encourage greener investment practices (cf. Kauko, 2011a). The final point made by Addae-Dapaah and colleagues is that 'the sustainability of sustainable property development' depends on quantifiable and realisable economic benefits. This is easy to agree on – credibility always matters in adopting/adapting new practices.

The study by Harrison and Seiler (2011) on the role of political ideology in the (rental) valuation of environmental certified commercial property in the US is particularly interesting in this genre. Using a database of 20, 172 industrial properties and OLS regression they find that 'green' certification increases the rents 8% in liberal (Blue) areas but decreases the rents in conservative (Red) areas compared to a standard building (in this case industrial warehouse). They control for several factors of related to both the property itself and the area where it is located. Thus the conclusion is that Red and Blue counties have different behavioural tendencies and incentives schemes. Because of this finding Harrison and Seiler advise extreme caution in generalising about effects of certification on price.

Finally, Rauterkus and Miller (2011) examine the impact of walkability to neighbourhood amenities on land values (so not values of built property). Walkability is measured as a 'Walk Score': a straight line distance measure for the amenities situated within a one mile radius. High values indicate 'Walkers Paradise' and low values 'Car Dependence'. Here the idea is that value is added by the neighbourhood. Other factors that affect price as controlled for in OLS regressions. They find that walkability (perhaps as expected) generally increases land values but that this impact reverses when neighbourhoods become car-dependent.

In sum, a rather wide range of estimates for rent and price have been reported (see also Warren-Myers, 2012). This suggests a methodological backwardness still prevails when this line of research is being defined. This is a young tradition that needs to find the benchmark empirical evidence and that is not yet the case. A not insignificant body of research has opted to use the hedonic approach and multiple regression analysis (MRA).Warren-Myers and Reed (2010) do not, however, see hedonic price modelling as a sufficient methodology to verify any existence of sustainability elements in commercial property values.

Here a caveat is in order. In much of the abovementioned discussion the issue is simply whether green elements will increase prices. The problem of hedonic and other similar studies aimed at isolating the price lift (or other hypothesised economic benefit) is that they belong to the same old paradigm which arguably has caused the problem. We have not yet seen the real pioneering contribution in relation to sustainability in this vein no matter what the authors listed earlier will claim. The difficult issue here is to track independencies across the sets of attributes rather than to isolate a single sustainability generating one. Nevertheless, a few other research communities here come to mind, perhaps most notably work by David Lorenz and Thomas Lützkendorf (Karlsruhe Institute of Technology), Georgia Warren-Myers (Deakin University Melbourne) and Eli Støa (NTNU Trondheim) are such promising endeavours.

Warren-Myers and Reed (2010) underscore the importance of having more transparency between sustainability and economic returns or values, their argument being about a negligible correlation between value and sustainability. They note that not only is this relationship subject to great uncertainty, but there is also uncertainty in the industry surrounding 'the accurate measurements of sustainability in reference to commercial real estate' (cf. Støa, 2009). To provide such info Warren-Myers and Reed (2010) carry out a survey of the opinions of investors and valuers about how sustainability is perceived in the industry by each actor group. They conclude that the difference in opinion between valuers and investors is due to the tendency of the former to be misinformed about the strategies of investors that focus on cost minimisation and value enhancement related to sustainability (cf. Kauko, 2011b).

The lack of data caused by infrequency of transactions and questionable level of 'hands on knowledge' about how sustainability influences value (i.e. by valuers) notwithstanding, Warren-Myers and Reed (2010) observe more than 600 rating tools all over the world. Most of them pertain to green building concerns, however, which suggests a discrepancy insofar as we are interested in sustainable land and buildings. The Green Star certification system in use in Australia, however, come closest to such a holistic system, they note. Even after the study by Warren-Myers and Reed the question remains as to what is 'truly sustainable real estate' and not just 'green buildings' in this context. Here it is important to note that all these results are about the greening of the dwelling and building, not the surrounding neighbourhood. (We return to this in the next section.)

Much has been said about the various rating systems. Goering (2009) points out that the US-based LEED system is one of the few holistic ones in this context (although as already mentioned, Jackson (2009) found LEED development options riskier than their Energy Star counterparts). He argues that is not enough with green building but ecologically framed site selection must always be considered first. For example, Jonathan Rose companies in New York City begin with choosing the right place to build before considering what to build. Energy efficiency is on the rise and affordable housing developers in particular seem to have internalised this. However, measurement and research on improvement of rents and values is missing, which makes it difficult to validate any particular guidelines concerning design features. There is furthermore a need for clearer incentives, more formalised building code requirements and more innovative educational and training programmes concerning sustainability, Goering purports.

Runde and Thoyre (2010) have already provided a variety of 'hands-on' guidelines that seem well-thought-out as they are based on the LEED context. Their overall argument is that for sustainability to have relevance in real estate valuation, it needs to be distinguished from being merely green; sustainability also encompasses issues of food, conservation and CSR, they note. Furthermore, green building criteria ought to be separated from 'green washing' which is not true sustainability at all – only claims about such. They even go as far as proposing an innovative methodology for approaching the valuation process with respect to sustainability/green issues, as they note that no such system currently is in (at least wider) use. Such a system ought to be flexible as it allows for market changes, they rightly argue. Like Addae-Dapaah and colleagues (2009), Runde and Thoyre speculate about the possibility of internalising negative externalities such as greenhouse gas emissions, traffic burden and infrastructure in the form of carbon taxes, special assessments and impact fees.

For valuation, Runde and Thoyre (2010) offer a relevant and reasonable proposition: that green building must meet three criteria: (1) commonly accepted set of features, (2) independently verifiable features, and (3) verifiable performance. Runde and Thoyre also make the important point that rating systems are not equal – some (e.g. Energy Star) limit themselves to energy efficiency, whereas others such as LEED take into account other features too. Furthermore, these authors assert that to establish a sustainability oriented market, the sustainability and green building policy ought to be as localised as possible. They set up clear and innovative guidelines about the valuation of sustainability related property features in not sustainability oriented (NSO) and sustainability oriented (SO) markets; the idea is that in an NSO market the property valuation outcome requires further analysis depending on if the case represents a 'brown' (i.e. not green enough) or green property. In the former case this situation might lead to a value discount, whereas in the latter situation it might lead to a corresponding premium. In the SO market context on the other hand the matter about premium (for green) or discount (for 'brown') matter is more straightforward, these authors suggest. Runde and Thoyre conclude that scrutinising sustainability affects everything around us – the unsustainable properties too.

Other studies include Lorenz and colleagues (2007) on residential property; Ellison and colleagues (2007) as well as Sayce and colleagues (2007) on commercial property; and Hemphill and colleagues (2004a, b) on urban renewal projects. Furthermore, less academic but related material has been published by Cox and colleagues (2002), Taylor Wessing (2009), RICS (2010) and Sayce and colleagues (2010). The practically relevant message conveyed by all this literature is that whether green or social sustainability features (e.g. innovative controls of either low or hi-tech type) can be seen as a feasible real estate development/investment strategy is an open question at the moment. The conclusion of reading the literature is that the benefits of green building are unclear; furthermore, there is a need to switch the foci towards the social and economic attributes of the location rather than looking only at green building, but that it is difficult in the absence of tangible and agreed upon sustainability principles – not to mention the absence of apt metrics.

The economic and social sustainability of urban locations

Sustainable location choice

Hedonic and buyer's choice studies alike indicate that a real estate market sustainability strategy is not only about the role of the buildings and built structures, but also about the role of the location in such strategies. In fact, the location share of real estate is after all at least as important a determinant of investment and value as the building on balance – although (as noted earlier in Chapter 2) many opposite cases of course exist too (Kauko, 2006b). The role of location and neighbourhood is, however, different from the role of the building. Logically, the quality of the home interiors are pertinent only to the residents living there, or those potentially living there in the future, whereas the quality of the façades, street, block, vicinity and neighbourhood is subject to evaluation also by neighbours, passers-by and visitors to the neighbourhood. If the house is situated in a city-core neighbourhood, the level of its quality and maintenance affects the image and attractiveness of the whole city too (see e.g. D'Arcy and Keogh, 1998; Musterd and Deurloo, 2005; Schwegler, 2006). Besides the overwhelming importance of the various elements of location from an essentially static perspective, of the two elements the location is more difficult to alter, and in doing so government cooperation is required. In this way we can see the importance of attractiveness analysis at a well-defined residential area level.

Zuindeau (2006) asks whether taking the spatial dimension – and not only the intergenerational one as is usual – into account changes the challenges of equity and efficacy, associated with the sustainable development issue.[11] He points out the problem that the spatial distribution of sustainability hitherto has been a neglected aspect of sustainability. The size of the area is an important parameter: in small areas the polluter-pays-principle is applicable, but when the area size grows the uncertainty about effects increases. Thus the real problem is the different policies implemented in relation to sustainability of adjacent territories that belong to the same higher level territory. He concludes that the local Agenda 2, as the institutional approach to promote sustainable development in towns and regions, faces a paradoxical problem: it is supposed to trigger the governance towards a favourable situation with respect to equity and cooperation, but the reality involves restrictions for such gains in the form of inequality and competition (cf. ESPON, 2010). Much related to this, Seyfang (2006) is concerned about achieving sustainable consumption in a situation when the current socio-economic frameworks put limits to the changes in consumption behaviour of individuals; according to this study, the following five indicators could set the new agenda: localisation, ecological footprints, community building, collective action and building new social institutions (cf. Munday and Roberts, 2006).

Even if we restrict the perspective to an individual buyer's/household's choice, (re)location cannot be overlooked in the analysis of residential real estate market sustainability. Koopman (2008) refers to Schelling's segregation model based on a two-dimensional grid, and argues that poorer enclaves in wealthy neighbourhoods tend to be more sustainable than pockets of prosperity in poor neighbourhoods. This is because of the different scale of the impact areas of entry and exit: namely, the prospective in-movers consider a wider area than their out-mover counterparts. In other words, when the experience/information about the location (i.e. site) becomes unfavourable it triggers out-moves more than what a favourable location would trigger in-moves, because, only in the latter case, the movers are unaware of the actual drawbacks and benefits of the site and have to rely on hearsay about the whole area (i.e. neighbourhood).

This can be explained more formally as follows: assume a location (site) A situated within an area (neighbourhood) B, a group of potentially out-moving habitants of A and a group of prospective in-movers to A. Furthermore, when deciding on in- and out-moves, all other influences than individual preferences for the site and neighbourhood: that is the experience/

information of A and the reputation of B, is assumed away. In such a situation the out-movers from A consider both the experience/information of A (which they know) as well as the reputation of B (which they may or may not know); the majority of the in-movers, who are not that familiar with A, in turn consider only the reputation of B and assume the reputation of A to be the same. The difference is that the out-movers already have lived in A and have experience/information which the in-movers do not have, even if the latter group have carried out surveys and may have even visited the location. Naturally, when aggregating this moving activity the net of attributes of B obtains a bigger weight than that of A, which then needs to be taken into consideration when determining the criterion for sustainable moving. Thus, assuming that price correlates with social status (as is reasonable to assume after reading the micro based NCE literature), moving to a location (i.e. site A within area B) where the average price level of B exceeds that of A is sustainable, whereas the opposite case, the price level of A exceeding that of B, is not. Along the lines of this analysis, Koopman concludes that the Dutch policy of mixing, namely to attract high income households to low-income neighbourhoods, is less effective than its American counterpart, the Section 8 programme, where the idea is the opposite: to subsidise the entry of poor households to wealthy neighbourhoods (cf. Cameron, 2006; Meen and Meen, 2003; Fossett and Waren, 2005). This issue is furthermore politically toxic, one could add here.

Urban regeneration and sustainability

As already explained, the government has an active role in shaping the location, which may be carried out either in a more traditional manner: by providing on-site and off-site infrastructure such as stops for public transportation, parks and recreational areas or improved safety measures; or in accord with more modern ideas of image creation and territorial competition policy. When considering this role of government more closely, two opposite ideals can be debated (see Dixon et al., 2005, pp. 19–24). The first is a neo-liberal policy of letting individual location choices determine the city structure, which aggregates to a spatially de-concentrated real estate market. Here the problem will inevitably be the resulting environmental externalities, notably environmental hazards[12] and urban sprawl. To avoid such categorically unsustainable outcomes another ideal has been given attention: policy making inspired by the political economy and regulation school traditions, according to which urban restructuring carried out by an educated local government is the solution to all spatial problems arising from an unhinged market of space. This in turn aggregates to a more consolidated land use pattern and a more city-like real estate market character.

However, while it principally is beneficial to develop an urban area as dense as possible, it is bad if an urban regeneration effort does not correspond with consumer preferences for such living, or if the city centre does not provide employment or leisure opportunities. This trade-off between two spatial ideals would suggest an 'in-between strategy' of empirically testable research where QOL, price and various other socio-economic indicators are related to each other.

However, Percy (2003, pp. 207–208) notes that the sustainable development discourse much depends on the city context; according to her there is on one hand 'widespread support for the principles of sustainable development to underpin and shape urban development and regeneration initiatives', but on the other hand there are huge differences in the way the sustainable urban regeneration agenda is implemented: in some cities the approach is proactive, in others mere 'lip service' is paid, and in yet others initiatives are created. Elsewhere, Bramley and Power (2009) emphasise the trade-off in sustainability between functional (i.e. access to services) and social dimensions.

The literature suggests that the social dimension of sustainability is still under-researched in relation to an urban regeneration context. Social sustainability is still a fuzzy and theoretically and methodologically limited concept subject to fragmented approaches. While many different traditions of urban regeneration have covered a social view, any particular approach (property-led, business-driven, urban form and design perspective, cultural industry, health and well-being or community-based social economy) has lacked an integrated approach, Colantonio and Dixon (2011) argue, before they set their own agenda with the following issues themselves:

- Inclusion of newer and 'softer' themes such as happiness, social mixing, sense of place, participation, needs and social capital to blend with traditional policy areas such as equity and health indicators.
- Community is not to be perceived as a monolithic block.
- Urban sustainability (or sustainable urban development) is part and parcel of a planning and urban regeneration policy of creating 'strong communities' even though 'the speculative nature of social sciences' remains (p. 35).

Dixon (2007) raises a number of points on what constitutes (un)sustainable brownfield generation in the UK in terms of the three pillars (i.e. the triple bottom line approach): economic, environmental and social sustainability. He also notes its alternative – the Russian doll model, according to which one dimension, usually economic capital, is considered the engine of everything and thereby more important than the other dimensions. Furthermore, when the economic pillar is examined, three factors become relevant: infrastructure, density and image, respectively, he argues. Dixon also investigates the 'policy push' and 'opportunity pull' forces of sustainable brownfield regeneration. A particular problem with the UK context is, however, the discrepancy between local effects and the regionally and nationally set planning regulations and delivery targets. Waterfront redevelopment is mentioned as a special case. Curiously the conclusion of his interviews is that, of two different kinds of risks, the contamination risk is less of a problem than the funding risk in sustainable brownfield regeneration projects. Implicitly, the institutional element stands up here.

While inherently unobservable and omnipresent, property market institutions are vital factors related to the provision and attractiveness of urban regeneration areas. To give an example, Ribeiro (2008) concludes, based on his undertakings on Lisbon's old downtown, that the public interest in physical restoration justifies major urban renewal projects, which in turn attracts private sector capital, leading to subsequent benefits in economic as well as social spheres. How then to evaluate the institutional structures and processes affecting neighbourhood revitalisation? For example, according to Schwegler's (2006) experience of Komárno (small town in Slovakia at the Hungarian border) the urban renewal was mostly successful: using a PPP arrangement and urban heritage marketing under-utilised areas in the inner city became revitalised, but at the same time, fragmentation of the townscape together with an emerging socio-economic stratification occurred.[13]

Elsewhere, Mace and colleagues (2007) argue that regional concerns may not fit a rationally set one-size-fits-all urban regeneration strategy, such as the one triggered by the urban renaissance agenda in England. Results from their case study of Eastern parts of Manchester indicate that, instead of implementing the urban renaissance concept, lower density brownfield development strategies can compete with suburban dwellings. This study suggests that, insofar as the goal is repairing of the perforated inner suburban fabric, a pragmatic agenda such as the one applied in their case study area might serve better than a compact city approach. Particularly as the latter strategy is unable to turn the push and pull factors towards sprawling suburbanisation anyway.

The issues involved are complex indeed. For instance, Taşan-Kok (2010) argues that a property-led urban regeneration project can be successful with respect to some criteria but unsuccessful in others, which she shows using case studies from two cities that are undergoing such regeneration: Antwerp and Rotterdam – in both cases the old port-areas. The issues involved are complex: global market is important but so is the local market; also the balance of private and public varies between such projects, as does the involvement of national and local government due to different macro-level institutional frameworks and governance cultures. For instance, in the Netherlands the national government has the most power in land use issues, whereas in the Belgian circumstances the local government is the most important tier.

Ganser and Williams (2007) remain sceptical about the suggestions that a national strategy for inducing brownfield development would lead to higher building densities and a more compact urban land use. Based on a comparison of how brownfield developments meet the targets in England and Germany, they conclude that this problem area involves too much ambiguity and one-size-fits-all thinking for a convincing claim to be made about the extent to which including urban brownfield development in the planning policy leads to sustainable cities. While the targets of 60% of the housing being built on previously developed land in England are, unlike Germany's 30ha/day target in all development, at least met in the planning goals, the problem is that reducing greenfield developments does not lead to a switch to more brownfield developments but to an overall reduction of housing developments, which consequently leads to price increases – thus an economically unsustainable outcome. It is furthermore a problematic assumption that that higher density new building on brownfield land is sustainable, insofar as it is a recognised principle that refurbishment is more sustainable than new development. Rather than confirming anything, the results of Ganser and Williams trigger new, more qualitative questions: is it *not* true that housing development in general is less sustainable than mixed uses, and that land use intensification is more sustainable than certain sustainable solutions (e.g. rainwater collection) allowed by lower density building only?

Another example: tax incentives can be applied to trigger more affordable housing/office package and a denser and more efficient land use, which in turn would lead to sustainable development. (Although see the previous discussion about the limits of the belief in increased density.) For a discussion on the extent as to whether the Lithuanian land tax system could be improved so as to embrace this aspect of sustainable urban development, see Raslanas and colleagues (2010). (In themselves, these issues of affordability and density are not against NCE theory either.) The main benefits of this proposal would, however, imply a number of sustainability aspects, such as more dense and rapid building of land, encouragement of owners to renovate or demolish shabby and unkempt buildings, more efficient and economical use of land, slowing down of road construction with energy savings and preservation of nature and landscapes as a result and price increases (see Raslanas et al., 2010).

Developing a methodology for evaluation of market sustainability

The previous discussion shows that real estate and housing market sustainability is an elusive criterion (see also Jones and Watkins, 1996, who launched this concept before sustainability became a popular theme). The core of the argument lies in the definition of *sustainable local property markets*. The other shortcoming noted was that, for this inquiry, certain *indicators* that enable analyses on market sustainability need to be identified. And following the general argument of this volume put forward in Chapter 1, sustainability involves much more than the 'green' issue, and that when measuring such factors, we can create an added value for real estate analysis. Could we utilise the possibilities offered by house price data for this purpose? Such an approach is sketched as follows. The proposed methodology deals with recording and classifying different trajectories of housing market sustainability for neighbourhoods and house types using data on urban house prices.

When different indicators are considered it should be stressed that we must operate on the best possible data we get, while at the same time acknowledging limits in comparability and compatibility. Earlier it was noted that two kinds of geographically identifiable data is required: property values and property descriptors. Valid property value data are easy to find in some countries and difficult/impossible in others. For example, in Italy such registers do not exist; in Norway such data exist in large quantities, but the majority of it is unreliable due to limitations in the register system; in the Netherlands it depends on the municipality; in England and Wales such data have only recently become available, whereas data were previously available only in Scotland or Northern Ireland. Other kinds of data should cover differences in the socio-economic aspect and differences in environmental/QOL aspects – note that pollution and other serious environmental problems in Eastern Europe pose a particularly strong challenge.

Data on environmental features, buildings and prices are by no means always possible to find (this is the case in Ukraine, for example, see Kryvobokov, 2004).[14] It may furthermore be observed that behavioural characteristics that define 'satisfaction' give an added value to the analysis (cf. the respective studies by Kauko, 2008a, Keskin, 2008 and Koopman, 2008). The data can for example be from housing demand surveys. Finally it is worth noting that the nature of the data in terms of its time scale and level of detail also depends on whether we opt for a predominantly neoclassical/quantitative approach or a more institutional/qualitative approach in the spirit of Vatn (2005). In both cases it should be a long enough time-period to register changes in physical sustainability dimensions: ten to fifteen years at least, but preferably more than twenty years.

To add a last few remarks, as the economic investment activity (perhaps worldwide) is expected to go up in the foreseeable future, an opportunity to get those investments on the right sustainability track is likely to open up. However, location matters alongside the building for the real estate investment and management. Here two strategies for creating sustainable markets can be outlined: one is 'elitist' and attempts to mobilise the behaviour of actors, with the role of government in providing stimuli (i.e. incentives); the other is 'cynic' and strictly focuses on the necessity for changing government regulation and new laws. Especially in the former case, but also when designing an in-between research strategy, the role of micro-level data analysis is strong: more particularly, the price in relation to quality or affordability. When this data are subsequently aggregated at a small spatial scale, it is possible to distinguish between trends that are completely 'sustainable', 'economically sustainable' or 'unsustainable'.

When analysing the role of location, it is crucial to apply an appropriate scale for the geographical or administrative area. Furthermore, difference across property types must be taken into consideration too. In fact, there is reason to believe there are even differences between sectors with respect to the correct 'economic sustainability criteria'. For example, if we extend the analysis into the office market, the operational criteria will be completely different due to fundamental differences in valuation method – income capitalisation instead of market sales. Such operational level differences notwithstanding, the same philosophy of sustainable components of property prevails on a conceptual level: for instance, a sustainable living environment means, among others, 'healthy apartments', and a sustainable working environment, correspondingly, 'healthy office space'. Thus healthy space has to be approximated into the particular criteria adopted.

Here several topics can be picked up for economic sustainability analysis. Following Kauko (2009a), in the present compilation of studies the empirical focus is on the quality related price (Chapter 6) and on the affordability of prices (Chapter 5). A third possible line of empirical modelling would concern the diversity of the product (see Kauko, 2009b, c). Using all this as a backdrop, the next chapter reviews some methodological and data issues that are vital for the analysis. It is to note that hereafter the text concerns residential property only, as there is more literature to build the argument around than other property and land use types.

Notes

1 One could add a fourth dimension here, namely the cultural one, and often it is indeed added in the form of 'value of something old and appreciated if not entirely functional' – either aesthetic or antiquarian value, or tourist attraction (see e.g. Kauko, 2013). However, from a point of view of measurability this side of the story is largely left aside here.
2 It is suggested that the cost of green building adds on average 2% to total development costs compared to standard buildings (Goering, 2009, p. 184).
3 The definitions used in the present book are different from the ones used by Bramley and Power (2009), insofar as the economic dimension is defined to include functionality criteria too (i.e. what Bramley and Power term social sustainability criteria).

4 Van der Maaten (2010) applies real option valuation in order to relate the investor's risk when investing in energy efficient buildings to compensation and subsidies. The idea of this theoretical framework is that when technology advances it is important to improve the payoff. This in turn necessitates investigation into subsidies.

5 See Wallner et al. (1996), who argue that 'islands of sustainability' – an island being 'an area where sustainability is reached at a local or regional level' – can act as cells of development within this context.

6 The Cradle to Cradle (C2C) approach is a refined version of this thinking regarding area development and building. The avant-garde urbanists of 'Lateral Thinking Factory' (e.g. Steven Beckers and Michael Moradiellos) argue that 'happy healthy cities' is a more apt concept than 'urban sustainability', as the latter concept [*sic*] still has some echoes of a more hierarchical past of failed policies. The new urbanist paradigm would be based on a cherry tree metaphor, where *effective* production would be the key instead of mere *efficient* production. In other words, while the cherry tree is rather inefficient in producing offspring, it is very efficient in producing hummus and CO_2. Apparently, Rabobank in the Netherlands and a number of financial institutions around the globe have already begun following such principles. Their motto is: 'don't take from the nature, borrow from it.' However, that all would be possible to achieve based on networks of private sector and NGO actors seems rather unrealistic at the moment, but in a more distant future this idea might well be relevant. Within this context of bottom-up thinking, *Holzmarkt* in Berlin, with its alternative and community oriented, eco-friendly and economically sensible business idea, is a fairly amazing project (see Holzmarkt.com).

7 A panel discussion workshop on sustainable real estate investment and management was arranged in connection with the European Sustainable Energy week 2008 (*Investing in a Sustainable Built Environment. Do energy efficient buildings make economic sense?* Wednesday, 30 January 2008, Brussels). One of the panellists, Dr. Angus McIntosh (King Sturge), argued that markets do not react to sustainability unless the law changes; thus only strict regulation can generate sustainability within the real estate market. As I pointed out in my response from the floor, this is a cynical point – even though McIntosh certainly is at least half right.

8 European Property Federation is in fact very much interested in sustainable property investment strategies and valuation methodologies. On a more informal plane, currently cross-national and cross-disciplinary efforts are also being set up, much at the initiative of RICS.

9 For a Real Option approach to analysing the viability of the *New Songdo City* project, see Geltner and de Neufville (2012a, b).

10 Arguably, NCE and the sustainability paradigm can be seen as incompatible insofar as green issues by definition require a longer term approach than the period of recorded data in the study by Fuerst and McAllister. Indeed, we are reminded of Albert Einstein's paraphrased wisdom : a problem cannot be solved based on the same body of knowledge that generated it.

11 Zuindeau uses the definition of weak (cost-benefit analysis) and strong (non-monetary) forms of sustainability.

12 According to Andrews (2008) governments should strictly regulate the development of coast locations. He argues that environmental hazards become a social problem too as there is unequal exposure to weather risks across social groups in relation to power.

13 Having recently had the opportunity to visit this town, my (obviously subjective) evaluation is that the outcome cannot be considered comprehensive or just – and certainly not sustainable (or even efficient), even if PPP has been applied for the projects.

14 Apparently, many aspects of the property market and land use were more sustainable during communism in Eastern Europe. Marko Kryvobokov points out that, for example, in Donetsk, Ukraine water was drinkable (but not anymore), accommodation was constructed in big volumes and distributed from free for employees of state enterprises and so forth whereas now there are low pensions, no medical insurance, implementation of anti-ecological policy, increased pollution and so forth.

4   Some diagnostic issues

According to the 'backbone' of this volume, sustainability is more than only 'green' features, in general, and also in a specialist context such as real estate. The rationale here is that social and economic dimensions, including the market value of 'green' attributes, need to be investigated in order to identify and assess the relevant socio-economic issues which, despite their relevance in everyday real estate jargon of professionals, are neglected in the literature – a point made by Lützkendorf and Lorenz (2014). And when carrying out this inquiry, a measurable approach is recommended – in particular the investigation should focus on how *price* can be brought in. Here a spatial approach is preferable and alternative approaches are accepted too. When all this is taken into consideration, and such metrics is developed, the sustainable real estate analysis obtains added value in the sense of improving decision making at the marketplace.

To give an example of the problem, let us retrieve the main theme of Chapter 2. The housing bubble, and its consequence, the credit crunch, are familiar terms now. The underlying reason for this phenomenon is said to be an unsustainable property/housing market. Unsustainable means 'not sustainable'. What then would be a 'sustainable market'? To find out, let us analyse the meaning of the concepts 'sustainable' and 'market'.

To start with the 'market', it is an economic concept, and pertains to a process where the driving goals of its participants are in one way or another aimed at reaping a profit as a producer or intermediary, or maximising economic utility as a consumer. Profit (and indeed utility), however, refers to a short-term gain; furthermore, it is relatively one-dimensional, even if there are schools of thought that do not refer to 'maximising' but merely 'satisficing' strategies (notably, Thorstein Veblen's 'old' institutionalism and Herbert Simon's behavioural science). These are in fact the key differences to the concept of 'sustainable market': its goals are defined as long term and multidimensional (Kauko, 2008a). Sustainability is multidimensional even if we are restricted to only the economic dimension. The previous chapters touched on these issues, and the present chapter documents further progress in this project based on desk research, observation and roundtable discussions. Here some cautious speculation is in order. In particular, how to define a sustainable market for operational purposes? The overview of the text is as follows: (1) further argumentation in relation to sustainable real estate, (2) design of methodology to sort out the conditions and actions for sustainable urban real estate, (3) review of data issues that would support the suggested methodology, and (4) summary of the chapter.

Lessons learnt

Thus far it was established that strategies for creating a sustainable market ought to be based on the following principles:

- The immediate aims likely to be 'bottom-up' (as opposed to top-down, i.e. the C2C concept brought up in an endnote in Chapter 3 in this volume).
- To convince investors, developers and homebuyers about sustainable market strategies.
- A role for government in stimulating conversions, refurbishments and new developments into sustainable modes.
- 'Elitist view' or 'cynical view'? Two strategies for creating sustainable real estate locations can be outlined as follows:

  1 *Elitist*: market behaviour of innovative actors and government stimuli (including Smart City and New Public Management enthusiasts).
  2 *Cynical*: government regulation and laws together with inclusive cooperation (including critics of Smart City and Communicative Planning Theory enthusiasts).

- Sustainability incentives such as taxes would be in between 'Elitists' and 'Cynicists'; thus convergence between the two ideals was noted.
- This is not only about the role of the buildings but also about the role of the location and neighbourhood features.

A number of caveats and qualifications regarding the previous definitions and the assumptions of the analysis need to be reflected upon. To start with, when propagating the 'elitist' stance it should be noted that certain basic regulations in relation to social and land use issues are needed as well. Without them no market sustainability can be established (e.g. Reese and Sands, 2007). For example, in New Zealand the spatial building and development policy was deregulated completely and the resulting neo-liberal policy led to a pattern of densities that surely was too low and thereby considered unsustainable.[1] Subsequently, the policy changed towards reregulating the land use. Here it is to observe that the change in density is not a straight-forward approximation of economic sustainability – namely, because the new stock is never exactly comparable with the old stock (cf. Kauko, 2003; Cheshire, 2005).

Residential density and planning regulations may well be overvalued in this context. Jones and colleagues (2009) in fact estimate the viability of housing developments based on model estimations in five British cities (i.e. Oxford, Leicester, Sheffield, Edinburgh and Glasgow) and conclude that the differences are primarily caused by to socio-economic factors and not density or house type regulations set by the compact cities agenda. In other words, that the existing constraints for viable housing development and sus-tainable urban form vary between areas across and within cities in the UK is not so much connected to the costs or risks of brownfield development as to the affluence of the given city and neighbourhood. In any case, the study of Jones et al. shows that a need for trans-disciplinarity is reality – that is the necessity to allow the participation of non-academics and non-experts into the academic or expert discussion. Hobson (2008) argues that 'ethics' and 'participating' provide important views for developing the sustainabil-ity discourse. It should be agreeable in general that knowledge generated outside academia should be recognised and given credit too. Visiting EXPO REAL 2008 in Munich – an essentially non-academic event of gargantuan proportions – gave me the picture that the ideas of sustainable real estate are rather well received in business realms. Furthermore, according to Judi See-bus and other panellists, sustainability always is a local issue; and impor-tantly, metrics of sustainability is a critical issue for investors (Sayce et al., 2007). The panellists stressed the need to adjust the sustainability in relation to the market. On the basis of this experience it is safe to assume that the ideas of sustainable real estate are much better accepted and developed in business than in academia. A big discovery (from EXPO REAL 2008 in Munich) was that the real innovative persons in sustainable urban develop-ment and beyond can be the private sector property investors rather than academics.

Among the benefits of attending such meetings are the possibilities to learn about various private investment and city marketing perspectives applied in practice. An example of borderline between practice and research concerns the issue of active vs. passive sustainable developments. Accord-ing to the panellists (Paul Appleby, Peter Höppe, David Lawrence and Mark Faithful) the problem is that the owner does not benefit of energy efficiency. Therefore, in Germany an energy passport (documentation of daylight, floor-tiles etc.) is required of each house by law. Unfortunately the problem often is how to oblige investors to educate better engineers. This is the 'active' sustainability school. As a contrast, the so-called 'passive schemes' – the ones which are connected to main networks – fit the bigger picture.

How then should sustainability be treated in relation to the financial crisis? As it is time to take stock and learn from the financial crisis, assets with energy certificate might become more widespread, these experts assert. The interesting notion is that most straightforward business people are convinced about sustainability as a criterion of investment and fully believe in the concept of *sustainable real estate*. To some extent professional organisations such as RICS believe too.[2] But the disappointing and frightening thing is that so many academics are not convinced. For many, the issue is about exact definitions in a situation where such luxury as sustainable real estate is yet to be established. Is it sustainable real estate or sustainable real estate markets? Why not include both concepts in the same framework as the market is always the aggregate of real estate sales? Logically, improving the quality of the real estate works the same way as improving its market position. For example, planting a tree – or even better: keeping some of the existing forest – not only leads to an increase in use value but also is bound to raise the market value and potentially realisable sales prices in the area.

In an attempt to reinforce our belief in public policy, environmental economist Bromley (2007) discusses market failure – a defining concept in the environmental economics paradigm – in designing sustainable environmental policy. He argues that 'sustainability concerns the creation of dynamic human processes'. Bromley acknowledges the key point of all sustainability arguments: 'the entire process of environmental policy is to make sure that future persons do not acquire the "wrong" preferences.' Furthermore, his claim that 'models of optimality bring nothing compelling and necessary to the realm of human action' is also correct – true that statistic and other overtly 'scientific' approaches have their limits. Some of his arguments are debatable, however. What he fails to note is that the sustainability goal also incorporates non-monetary costs. Then he mocks 'scientific approaches' and objective measurement of sustainability, which to me is rather misguided and cynical even. Finally, he suggests that sustainability is a democratic concept – this is something I disagree with, because professionals tend to know better problem solutions than lay people.

Bromley's arguments are of course presented within a 'relativist' perspective (cf. Hobson, 2008). Some of these arguments seem reasonable and innovative and others rather doubtful and even downright erroneous. In my view, designing democratic policies, Pareto optimal or not, are seldom sustainable simply because designing any policies is lagging behind the actual social and market-based processes. What seems a more agreeable proposition is to recognise that unsustainable real estate practices are based on too high expectations of profit (e.g. Wallace, 2008). A theoretical case can be made for two types of price setting situations: one based on a *real effect* of the housing market where the price changes with time and another one based on a *spurious effect* where only the stock changes but price per unit does not change (Prasad and Richards, 2008).

Moreover, sustainable property investment strategies may be connected to a property rating system that focuses on the property to be financed rather than on the credit standing of the borrower. The individual property asset based rating is a relatively new instrument, the cutting edge development and application taking place in the German speaking Europe and to some extent already in the UK. Their benefit is argued to be their ability to create opportunity and risk profiles of property assets. Due to their treatment of 'unsustainability' as additional risk element these rating systems can offer guidance about sustainability for banks and investors. As a consequence of such communication, a market transformation of the construction and property sectors will eventually occur. However, the challenge is to harness property values and the financial instruments to reflect the true market value of sustainable buildings – to demonstrate how such buildings perform under existing valuation and risk analysis methods. Moreover, social, psychological and health effects related to the surrounding environment, for instance, good landscape aesthetics and planting trees, or waterfront developments, will also be taken into consideration in such a methodology. More research is, however, yet needed on many of these issues, such as on the relationships between technical, functional, environmental and social performance and the economic effects (RICS, 2007).

A methodology for empirical analyses

An 'academic tradition' targets associations and correlations between property price and other physical, social-economic and environmental (and sometimes institutional and behavioural) variables. This research was largely established though the analytical tool of hedonic price/market modelling which was developed in the late 1960s to early 1970s, and then from the late 1970s onwards refined through the segmentation/submarket concept (e.g. Maclennan, 1977; Grigsby et al., 1987). On the other hand, a 'practical tradition' that is preoccupied with improving the modelling performance for mass appraisal purposes coexists (see Lentz and Wang, 1998; Kauko, 2004c, 2008a). However, various academic empirical traditions could benefit from sophisticated value modelling tools when applied as a guideline for which market model is valid and feasible for a given dataset with certain recognisable tendencies.

In contemporary methodological perspectives to house price analysis/ modelling the market is considered idiosyncratic with respect to one or more of its fundamentals, in which case the differences across locations and housing bundles are more qualitative than quantitative by nature. This way, the urban location can be analysed either using a simple equilibrium or a more context-specific model, based on multiple equilibria. There are, however, no clear guidelines for when to apply what kind of model/method. The questions to answer are how physical, socio-demographic, financial and administrative factors shape the housing choices of individual house- holds, and this way the urban form. While these questions are important, they are also complex; thus, if we want to perform empirical tests on these issues, we need simplifications.

The selection of modelling approach is the key to success when the cri- terion is 'realisticness' as opposed to formal elegance. This principle is not difficult to grasp: what kind of market – that kind of method (see d'Amato and Kauko, 2008). *The self-organising map* (*SOM, Kohonen Map*, see Figure 4.1) is one example of a promising approach within this realm of realistic but less elegant modelling techniques (there are several other com- parable techniques, for example genetic algorithms and case-based reason- ing, see Kauko and d'Amato, 2008). Using machine learning and adaptive computation jargon,[3] the SOM is a type of *unsupervised neural network* with *competitive network architecture*. The neural network is a sophisti- cated statistical method that captures non-linear but regular associations (i.e. patterns) within a dataset without a pre-defined model. The SOM is best defined as a mapping from a high-dimensional data space onto, a (usually) two-dimensional lattice of points. This way disordered information is pro- filed into visual patterns, forming a landscape of the phenomenon described by the dataset.

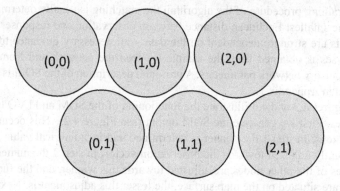

Figure 4.1 The situation of the nodes in a three-by-two (3 × 2) map.

Here the aim is to classify the citywide residential property market (or the owner occupied segment thereof). Using the definition of Kauko (2008a) a classification of housing market sustainability to various ordinal classes of sustainability would be a prerequisite to subsequent value modelling, in order to determine whether the long-term outcome is of a healthy or distorted kind. In the long term, the market can be classified, to various degrees of sustainable or unsustainable types, and this is not the same as classifying it as efficient or inefficient (although the two dimensions may overlap). From the concept of market sustainability we can then deduce the premises for value sustainability: a sustainable market generates sustainable value, which then can be used as an attractiveness indicator in a broader sense; or in the opposite case, an unsustainable market diagnoses a problem – an unsustainable value. This general model of the marketplace will subsequently be subject to empirical verification using sophisticated modelling techniques, namely the *self-organising map* (SOM, Kohonen Map) and its extension the *learning vector quantisation* (LVQ) classifier, on how data on the long-term development of prices and the quality of the residential environment (or a more general indicator, quality-of-life) are related to each other.

The overall principle of the learning process in the SOM is to 'train' the output lattice (map surface, feature map) based on the input data and some externally manipulated parameters in such a way that each original observation is matched with a unit (node, neuron) on the map surface where the numerical values measured in the dimensions of the observation are as close to those of the input observation as possible (see Figure 4.2). When this matching proceeds iteratively, we eventually obtain a projection of the input data. In this projection the topology between items is preserved rather well from the original data. The result is when the SOM, after a predefined amount of iterations, has then produced a feature map of nodes, each of which represents a characteristic combination of attribute levels. In the training procedure of the algorithm the matching is usually determined by the smallest Euclidean distance between observation and response. The results are strongly dependent on the data – all necessary guidance to the analyses is obtained from the sample we feed the network and from the compulsory network parameters. A profound description of the SOM is provided in Appendix 1.

Figures 4.2 and 4.3 illustrate the functioning of the SOM and LVQ techniques. First we generate the SOM output (see Figure 4.2). This occurs in two steps: in step 1 the winner is determined and its numerical value subsequently adjusted towards the observation vector; in step 2 the numerical values of the other nodes are adjusted towards this winner, and the further they are situated on the map surface, the lesser this adjustment is. Next we use the LVQ as a classifier. This classification is based on how we have

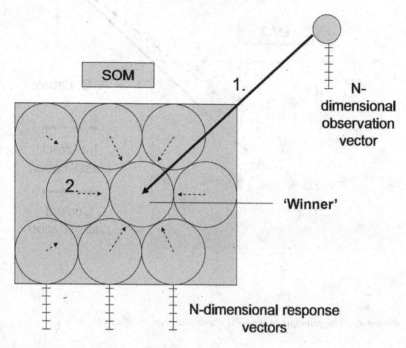

Figure 4.2 Illustration of the learning process of the SOM in two steps.

labelled the input (observations) and the response from the SOM output (see Figure 4.3). The result is a percentage correct classification of the whole sample plus for each category separately. We repeat these procedures for each year so as to obtain a time-trend.

The SOM only performs cross-sectional analysis – of price or any other variable or combination of variables. However, apart from the question of 'where?' the other interesting question to answer here is 'when?' Thus the answer cannot be provided merely by results of a 'model', that is one round of SOM computations. Because the objective of the present study is price movement, as opposed to cross-sectional price analysis, the SOM has to be used in a repetitive way. Carlson's (1998) *method of fixed time-windows* is suited for this purpose. It is a quasi-dynamic modelling device which can illustrate changes in the modelling outcome or in one particular variable such as price without using the time of sale as a variable. This is illustrated in Figure 4.4 (see also Kauko, 2009c). It is to note that the labels of the nodes are for identification only: each node with 'hits' obtains the most frequent label of the observations it 'won'.

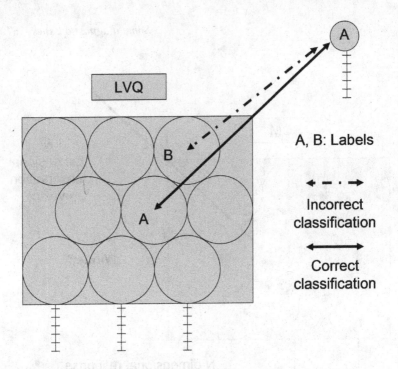

Figure 4.3 The principle of the LVQ classification of the dataset.

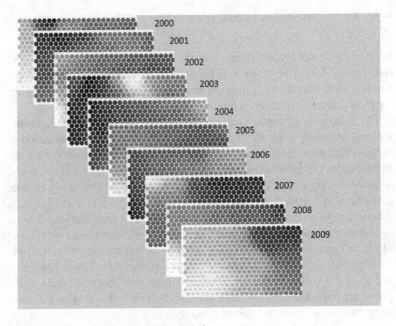

Figure 4.4 The method of fixed time-windows.

The results from a prior study serve to illustrate the possibilities (cf. prior analyses on Budapest housing markets by the same author, see Kauko, 2006a, 2009c). As shown in Figure 4.4 the SOM output identifies change in price level in Budapest, Hungary.

The empirical analyses

Research design

Given our overall aim of examining the economic and social aspects of sustainable real estate, ideal data for this study are data on house prices and various location specific features. As already discussed, it could, for instance, comprise citywide (administrative level of municipality) data on (1) actual transaction prices or hypothetical value estimates, (2) environmental and housing quality (or QOL), and (3) income and affordability. It is to note that, to analyse the aspects of real estate sustainability discussed in previous sections, we need data on property values and their descriptors; not only building data, but also about locations, that is to say, where certain types of sites are located. It can be argued that data on sales prices or assessed values of individual dwellings include plenty of useful information about the two key aspects of this study: about the attractiveness of locations on one hand (i.e. preferences) and the tightness of the market on the other (i.e. containment).

The empirical material of the study comprises data from two rather different urban contexts: Amsterdam, the capital city of the Netherlands; and Trondheim, a medium-sized city (that nevertheless is the third largest in Norway). Amsterdam is also undoubtedly a global city insofar as the city is distinguished by the strong presence of advanced consumer services (Kloosterman and Lambregts, 2007). Trondheim, in turn, can be considered a relatively provincial and peripheral city, but also one where the economic fortunes to a great extent depend on the high-tech and R&D sectors. These analyses are documented in two subsequent chapters of this volume.

Trondheim

In the Trondheim study the variable of interest is economic sustainability of real estate price development approximated through the affordability indicator: transaction prices divided by income. How are data on the long-term development of prices and affordability indicators related to each other? This is a yet relatively unexplored line of research.

The aim here is to analyse this sequence of cross sections using the SOM, more particularly the method of fixed time-windows developed by Carlson (1998). By examining successive feature maps of cross sections from 1993 to 2007 the price-to-income signals are seen as a trend. This way it can be established which cases have become the most expensive ones in relative terms, using location and property descriptors as identification.

When such a piece of analysis is carried out on a year-by-year cross-sectional basis it can be seen which house types and locations are more and which are less sustainable in terms of the upwards (i.e. decreased sustainability) and downwards (i.e. increased sustainability) signals when measured through the price/income indicator identifiable from the yearly SOM outputs. This is a pragmatic approach in the sense that we look for a locally adequate viable solution for a problem without need for general laws.

Trondheim is a mono-centric city with a small historic centre, relatively compact inner suburbs, and then a chain of sprawling outer suburbs, punctuated by agricultural land and forest. The terrain is hilly. This city kind of 'melts' into the countryside and lots of the suburban areas are characterised by rural architecture and streetscape – even the more recently developed ones – despite the city housing a population of *c.* 170,000. In Trondheim, like in much of the rest of Norway, the geographical circumstances dictate a great deal about how the settlement and mobility patterns unfold. At least the Northern and oceanic situation together with high altitudes is traditionally seen as such determinants (Ahlmann, 1917). Moreover, when examining the urban restructuring contexts in Trondheim, we note the following (see Figure 4.5):

- Not really bad areas but the southern suburban housing estates such as Kattem and Flatåsen are worst in terms of housing market and socio-economic indicators.
- Of the inner areas Lademoen is traditionally worst off in social-economic terms (e.g. Brattbakk et al., 2000).
- Upper market areas exist in Byåsen, Lade and Singsaker.
- Neo-liberalist developments are scattered all over the otherwise rather picturesque and, from a physical point of view, inefficiently built historical inner areas, most notably the waterfront development Nedre Elvehavn on a former shipyard site.
- New urbanist developments exist too (notably, in Ilsvikøra, the first urban renewal area in the city).
- Gentrification is proceeding at a variable pace in various parts of the inner city:

  - Bakklandet, Møllenberg is already gentrified.
  - Ila is on its way to becoming gentrified.

In this country the condominium sector has existed since the early 1960s and expanded rapidly after condominium conversion of rental dwellings

Figure 4.5 Map of Trondheim.

became the major form of market growth since 1970 (Wessel, 2002). According to Brattbakk and Hansen (2004) the post-war large housing estates in Norway seem to offer satisfactory conditions both physically and socially in international comparison. Nevertheless, this segment represents the lower range within the Norwegian housing market; many housing estates are stigmatised and some have experienced factual deterioration. In general, Norwegians prefer lower density living, these authors conclude. (Figures 4.6 and 4.7 show examples of residential environments in Trondheim.)

Housing provision in Norway is traditionally dominated by market economy. Owner occupancy is the most common way of living with a portion of 63%, whereas share-owning co-ops (14%) and private rental (18%) are the other main forms of housing (the share of public housing is only 5%). The ownership and tenure-structure is rooted in the institutional decisions made in 1946 in the context of 'social democrat ownership country' aspirations, where a cooperative housing model and State Housing Bank was established to fund and subsidise homebuilding. However, changes in the system since the 1980s have created a market oriented housing system. At present national policy is focused on integration and reduction of social and ethnic housing market mediated segregation. While average housing

Figure 4.6 Suburban post-war neighbourhood in Trondheim.

Figure 4.7 In Trondheim high-rise blocks are not widespread but exist nevertheless.

standards are high in Norway, a number of housing problems exist in relation to affordability, ethnic segregation (but manly in Oslo), limited demand side subsidies, and lack of long-term physical qualities that is sustainability (Holt-Jensen, 2009).

Much of the sustainable urban development strategies in Norway are manifested through the *Framtidens byer* (Future towns) coordination programme, where Trondheim is included too. Here the state and the thirteen largest cities have agreed to cooperate to reduce greenhouse gas emissions and thereby contribute to the improvement of urban living environments (Regjeringen, 2011). Besides this Trondheim has locally (i.e. at the municipal and county levels) articulated goals of sustainable development such as 'becoming a sustainable town in a sustainable region' by 2020 (Trondheim Kommune, 2010). National subsidies are aimed at building a better public transport system and more pedestrian and bicycle paths in Trondheim (Trondheim Kommune, 2012). Indeed, sustainability is strongly on the policy agendas. Whether the same can be said about the real estate development agenda is, however, less clear. While sustainability today in Norway too is seen as the ultimate goal, Medalen (2006) speculates about how densification of the house building in Trondheim might in the end lead to decay, and thereby not be helpful for generating sustainable results, despite such expectations. This would to a high extent depend on how low the share of rental apartments can be kept and how well a given development project can be organised, he argues.

Amsterdam

In the Amsterdam study which is follow-up to Kauko (2008a, b), the main question to answer is *Where?* The dataset on Amsterdam 1986–2002 is utilised as follows. Each observation is labelled into three ordinal categories with respect to all three variables: price related to dwelling quality, price related to maintenance and price related to quality of micro-location. Here it is to observe that, while all three are here treated as measures of economic sustainability, the last of the three could in fact pertain to social sustainability too – this was in fact the operational solution selected by Bramley and Power (2009), who themselves point out that they have no objection about disputes in terminology concerning 'sustainability' or 'social impacts'(p. 45). Here it is worth a reminder that these definitions are by no means standard and other authors have used different categorisations of what kind of indicator possibly could be termed social or economic (environmental dimensions being better agreed on). For instance Turcu (2009) defines housing affordability as a social sustainability measure.

Because of the relevance for the interpretation of the findings in subsequent Chapters 6 and 7, a few words about the Dutch urban context and the national programme for urban restructuring are necessary. In 1995 fifteen large cities and the national government signed a covenant, which was the basis for *the Big Cities Policy* (*Grote Steden Beleid, GSB*). The GSB resulted in an inventory of thirty cities (Grote 30 or G30). The GSB aims to improve the economic competitive power of cities and to restrict socioeconomic and ethnical divisions within cities. Subsequently, in year 2003 the Ministry of Housing, Spatial Planning and the Environment (VROM) formulated a *Program for Action Restructuring* (*Actieprogramma Herstructuering*). In this program fifty-six neighbourhoods (representing 219 districts) were assigned. This group of fifty-six included areas from the following Amsterdam *wijken* (the official district units in the Netherlands): Zuidoost, Westelijke Tuinsteden, Wijken 08, 09 and 10, Amsterdam Noord (see Figure 4.8). These neighbourhoods are assigned to improve the planning and negotiation processes between the various stakeholders and also to improve the vacancy chain on the housing market. This last aspect refers to the assumption that the primary strategic supply of new dwellings (with diversity in tenure, type and so forth) generates housing moves in the existing stock.

The increased residential mobility should lead to a different composition of the stock and households, and further to reduce the spatial segmentation, which in turn should facilitate bonding and bridging capital and investments. The justification of the selection of the abovementioned fifty-six urban regeneration areas is based on an elaboration about their social, demographic, economic and physical features. In general, these areas face various problems in the sense of high unemployment, large share of low-income groups and some more area specific features (e.g. high share of elderly or foreigners) that are supposedly causing problems in the long run, and thereby justify urban regeneration policy targeting. The majority of these areas comprise ethnic pre-war neighbourhoods in larger cities, but some exceptions include white working class neighbourhoods and post-war housing estates. The criteria for selecting them are the assumed urban restructuring problems in terms of economic, social and physical factors.[4]

In Amsterdam the share of owner-occupied housing is only *c.* 20% of the total housing stock. However, the remainder of the property stock also falls under obligatory tax assessment. This means that we have to interpolate and extrapolate from this figure when the whole Amsterdam housing market then is considered, which suggests that a measurement error is present in the majority of cases (the tax authorities, in particular, have this problem). Thus the values are not 'real values' in 80% of the urban area. Unfortunately this also applies for the present study which uses this data source! Even so

Figure 4.8 Map of Amsterdam.

a method based on transaction price data gives some hint of the economic sustainability, defined as value stability, of the residential real estate stock in Amsterdam.

Summary of the aims, assumptions for research on sustainability using the SOM

The overall theme of this volume is to create socio-economic sustainability metrics using data on locations and prices, including monetary value of environmental and ecological sustainability features, to aid real estate related decision making. When we consider how and by whom this is best done, an interesting state of affairs can be observed: apparently the ideas of sustainable real estate are much better accepted and developed in business than in academia. The statement made in this chapter is to look at practitioners rather than 'ivory tower theorists'. It can furthermore be argued that sustainability always is a local issue. Here a metrics of sustainability, in relation to the market, is a critical issue for investors. Sustainable property investment strategies may be connected to a property rating system that focuses on the property to be financed rather than the credit standing of the borrower. Social, psychological and health effects related to the surrounding environment also have to be taken into consideration. However, more research is yet needed on this topic, considering its importance in social-economic terms.

Elsewhere high theorists like Bromley (2007) have a belief in democratic public policy within this realm, and therefore consider sustainability a relative and rather subjective concept, thereby supporting the public policy and government regulation based 'route' to sustainability practices. This seems questionable given the historically poor capability of any hierarchical systems of provision to adapt. On the other hand, if more of an elitist stance is taken, the role of the marketplace takes the centre stage. Consequently, the role of micro-level data analysis is strong: the encouragement is to use an indication based on the price in relation to quality or affordability. This way it is possible to distinguish between trends that are completely 'sustainable', 'economically sustainable' or 'unsustainable'.

Against this background, the logical aim of the empirical part is to classify the citywide residential property market. The assumption is that a sustainable market generates an attractiveness indicator in a broader sense, or in the opposite case, an unsustainable market diagnoses a problem. In the analysis to follow, one dataset comprises transactions of (detached) single-family homes from Trondheim 1993–2008 with only a geographical location and plot size as variables. The other dataset in turn comprises transactions of dwellings from Amsterdam 1986–2002, with sales prices and a number of attributes including subjective environmental and housing quality as input variables (property descriptors). Subsequently sophisticated classification analysis is conducted in the next two chapters.[5]

Notes

1   Professor John Henneberry, University of Sheffield, informed me of this case in an informal discussion (Leuven, 7 April 2009).
2   RICS (2011) themselves state increasingly sustainable business practices, thereby leading by example.

3   A novel with ties to this topic could be mentioned in this vein: Robert Harris, *The Fear Index*. The plot is about a hedge fund manager who invents a method to beat the market using machine learning (or Artificial Intelligence) algorithms. This method is then put to use in short sales of futures and options. After a promising start, this has disastrous consequences for the global society even though a small group of shareholders gets richer. This kind of development could also happen in the real estate context.

4   There is follow-up to this policy known as *Vogelaarwijken*, following the new cabinet 2007 (and its Minister for Housing, Neighbourhoods and Integration in the Netherlands, Ella Vogelaar). These are supposed to be the forty worst *wijken* in the country. While the number forty is chosen rather *ad hoc*, the selection of these areas is based on a more quantitative evaluation of the physical and social transformation than the fifty-six of GSB.

5   Last, a methodological remark is in place. Countries where demand has risen strongly in recent decades due to increased incomes (e.g. Norway, the Netherlands) show a total negative impact on a sustainable market, as noted by Fotis Mouzakis (Cass Business School, RSA conference, Leuven, 7 April 2009). However, it is still possible to disaggregate the indicator according to functional and spatial variation.

5 Quasi-dynamic SOM analysis of affordability in Trondheim 1993–2007

Description

The observation about urban sustainability being much more than the 'green' building element forms a kind of 'backbone' for this book. While this state of affairs may seem obvious to many readers, the lamentable fact is that socio-economic factors, including property prices are, despite their importance, yet neglected in both academic and practical literatures dealing with sustainability and sustainable development. To improve the credibility of incorporating such elements, this book argues for the necessity of designing a socio-economic sustainability metrics for urban real estate. And in this vein it could be argued further that such an advancement of the literature would bring added value for the real estate analysis, particularly in the urban context, where the multitude of socio-economic factors is a vital issue to deal with. This would also likely trigger a further adjustment of the analysis towards recognising socio-cultural factors such as lifestyles and images, although, due to their predominantly qualitative nature, not within the same metrics as the socio-economic factors. Whereas the first four chapters of the book discussed the theoretical-methodological side of this argumentation, the next three chapters move to the empirical side.

The particular aim of the exercise documented in this chapter is to compare house prices with an affordability indicator constructed in simple terms, that is to say, income. Relevant information for the study is assembled from the Norwegian property registry on one hand and on the income statistics on the other hand (the income variable comprises median gross income for the municipality based on the taxation of residents over 17 years old). The administrative city area of Trondheim (see Figure 4.5, in the previous chapter) is isolated from the data, which results in 36,613 sales transactions for the years 1993–2007. These are the years we obtained aggregate income data for too. Suspicious cases such as those with prices lower than an *ad hoc* cut-off point of 150 000 kr (*c.* 18,000 EUR) or containing (other) clear errata as well as non-residential use (offices, warehouses, shops, farms, public buildings etc.) were subsequently removed from the dataset. The data cleansing operation did not, however, end here (i.e. in preparing a dataset for all dwelling formats).

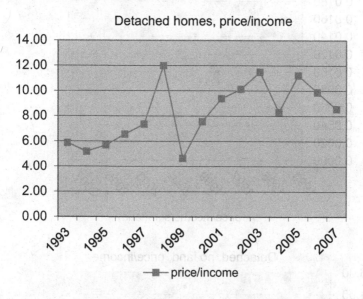

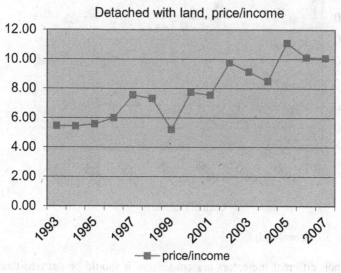

Figure 5.1 Development of prices in relation to incomes in Trondheim for the detached house type. All transactions; those without land included; those with land included; price per m².

(*Continued*)

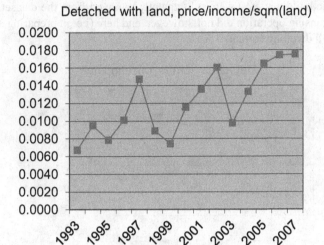

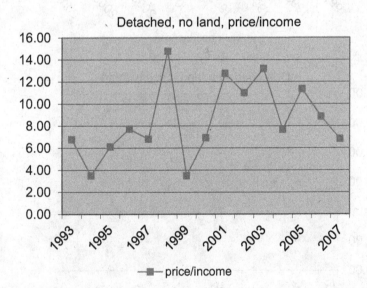

Figure 5.1 (Continued)

When different indicators are considered it should be stressed that we must operate on the best possible data we get, while at the same time acknowledging limits in comparability and compatibility between different datasets and variable definitions. Valid property value data are easy to find in some countries and difficult/impossible in others, as already discussed in Chapter 3. In Norway such data *does* exist in large quantities, but the majority of it is unreliable due to limitations in the register system: namely, the accurate information about the physical description of the dwelling unit sold is not recorded consistently. Hence, after a careful examining of the dataset it turned out that most of the data was not applicable, which was lamentable but not surprising. The data for other dwelling formats than the detached house type turned out to be unreliable as observations can be about several dwellings sold in one transaction without being coded comprehensively by the authorities (*Tinglysning*) who register this data from the transactions. Because of this building features could not be included as input variables in the analysis. In the end 5,260 transactions (between 183 and 735 observations for each cross section) were applicable for the analysis. The descriptive trend is shown in Figure 5.1. It is to note that, from an economic value stability point of view (see Kauko, 2008a, 2010), the graph shows unsustainable peaks for the respective years 1998, 2003 and 2005. These are visible – more or less – in all four graphs.

SOM analysis

In the outputs of the quasi-dynamic analysis with the SOM each feature map represents the transactions in one year. These can be analysed with respect to three direct dimensions: price, land area and location of the transaction (for identification only, smallest intra-urban ward/district level). Each feature map can also be analysed in terms of other, latent dimensions, such as topographic features of the terrain or special architectural/design characteristics of the house or neighbourhood. The map layer depicting the price of the property is picked for further scrutiny for all fifteen annual feature maps shown in the sequence of feature maps displayed in Appendix 2. Together these sixteen diagrams make up the time-windows application and are analysed as follows. (The corresponding diagrams for the other map layers are obtainable from the author upon request.) The same information would also be possible to show in numeric format – the graphic format is selected due to pedagogic reasons.

In most cases the same extremely high priced (i.e. from a value stability point of view) unsustainable area remains the same through time – it is about five or six adjacent neighbourhoods in Byåsen, the suburban district in the western part of the town (see Figure 4.5). Towards the end of the period the high priced areas are still identifiable also in other parts of the city: Jakobsli, a suburban location in the eastern part of town; as well as three locations in the inner city (or immediately adjacent to the inner city): Rosenborg east of the city core and Nedre Berg southeast of it, are the only locations that show up in at least two maps. It is to note that a single-family home in the inner city is obviously 'a rare butterfly'; for example, two suburban locations: Eberg, and Stokkan, both in the eastern part of town, show up in one map only, as do other outliers/anomalies. In such cases where it is obvious that the exceptionally high price of one property is due to a specific unrecorded reason; for example, in year 2005 a house in Lademoen, an otherwise unfashionable inner city area, sold for over 7 Mkr and the only recorded fact that supports the high price is that an area of *c.* 1,200 m² is included in the transaction. In other cases the computations have been robust, for example, in year 2003 Rosenborg represents by far the most expensive transactions and this case is actually backed up by 272 comparable properties (in fact more than every third observation in the cross section) that landed in the same node. That prices rose in Rosenborg is in fact contributed to the completion of a nearby shopping area and Waterfront development in Nedre Elvehavn, which improved both the status and accessibility of this neighbourhood.

Moreover, (apart from the Lademoen case mentioned previously) there is not much association/overlap between size of plot and price – especially not for the earlier years. The analysis reveals that the only continuous interval when the largest areas also happened to be relatively expensive properties was during the three years 2000 to 2002. However, these cases are never the most expensive ones during this period.

It is difficult to see any systematic for the data overall; some years certain locations are overpriced and some years others are; no connection with plot size can be confirmed; also not with location, beyond the tendency of being situated in Byåsen. Here we must add that the notable price premiums might be explained by the fact that Byåsen also has some of the best panorama views in town. Hence the SOM output ought to be supported by domain knowledge and expert information. Some of the findings are contributed to 'rational' neighbourhood effects, for example the abovementioned Rosenborg case, whereas other findings require more nuanced behavioural explanations about the long-term motives of the risk averse investors, who speculate about the future development of the area. However, the aim was only to demonstrate the usefulness of the method, rather than to delve further into the empirical findings.

We may now sum up and evaluate these analyses. Using data on property sales in the city of Trondheim, central Norway 1993–2007 together with a time-series on income aggregated on a city level from the same time-period, the idea is twofold: first, to look at a descriptive trend of price related to income to see how the citywide sustainability develops; second, to apply the SOM and the method of time-windows (Carlson, 1998). The output will here too indicate price/income. Here a caveat is to be noted: originally the idea was to accommodate district-wise aggregated income data of Trondheim. However, consistent income data aggregated on a detailed spatial level were not possible to get on a yearly basis, and therefore the method has to do with using the same income denominator for the whole dataset for each year. In other words, whereas the price variable is varying in time, space, and house-specific attributes, the income variable is only varying in time due to the abovementioned limits of this application. Nevertheless, this method is effective as each layer of examination shows the variation in price/income in a way that enables seeing the development of differences across locations and combinations of property attributes.

In line with the overall goal of this compilation: search for socio-economic sustainability metrics, this chapter demonstrated the possibilities afforded by a sophisticated approach based on property price data and other relevant variables. While the empirical findings have a value, also a few words on the method is necessary. The analysis of feature maps generated by the SOM has an intuitive appeal: non-linearity is allowed for, real relations between items are preserved despite a reduction of dimensions, and the outcome – patterns and clusters – can be examined both numerically and visually. The time-windows approach furthermore allows for some quasi-dynamics insofar as we treat the sequence of cross-sectional findings as time-varying dynamics. However, for a more rigorous classification this method is not suitable. Therefore another method, building on the SOM, is examined in the next chapter.

6 Quasi-dynamic SOM-LVQ analysis of quality in Amsterdam 1986–2002

Description

So far the story has been about socio-economic sustainability as a concept as well as a metrics. Continuing the task of demonstrating the value of numeric analysis, classification of property price data is propagated as an option in this chapter. For this purpose, a dataset comprising free market transactions of approximately 46,000 dwellings during 1986–2002 was prepared by the municipal tax authorities of Amsterdam (*Gemeentebelastingen Amsterdam*). This dataset is applied for the calculation of property taxes and contains recorded information on actual property sales prices and a number of attributes such as floor-space and subjective quality indicators (see Table 6.1). The latter are constructed by the assessor on an ordinal 10-point scale relating to the house, its maintenance and the micro-location (situation; immediate vicinity surrounding the property). Marks 1 (*very bad*) to 10 (*perfect*) are given for quality, maintenance and situation (0 is *empty* or *unknown*), respectively. This dataset can be conveniently disaggregated at district or house type levels.

Table 6.1 Individual data on market prices in Amsterdam, 1986–2002.

1	Transaction price and transaction price per square metre
2	Construction year
3	Type of house
4	House size (m²)
5–7	Marks 1 (*very bad*) to 10 (*perfect*) for other house-specific variables: quality, situation and maintenance (0 is *empty* or *unknown*)
8	Lot size (m² including garden, if any; in case of multi-storey apartment buildings, this indicates the size of the garden only)
9	Location by a canal
10	Date and year of transaction
11	Municipality land lease (*Erfpacht*)

The following needs to be noted:

- The identification is based on the name of the street in question, the district (*wijk*) and sub-district (*buurt*) and the specific format (*archetype*) of the house.
- The taxation dataset allows the use of price per square metre as a price variable, although the use of total price is a more common procedure in the Netherlands (in both academia and practice).[1]
- The *Erfpacht* variable indicates a favourable land lease contract and often a reduced price. These contracts ceased to be renewed after 1 January 2000, which supposedly have brought strong price increases (up to 50%) in certain cases.

At this 'first stage', a descriptive time-series analysis of the data was performed. Here price was related to each of the three quality variables. If this ratio increases, it indicates general unsustainability (using the definitions for this purpose: focus on economic and social issues); if it stays constant, it indicates economic sustainability (from a value stability point of view, see Kauko, 2008a, 2010), as the trends of both price and quality indicators are synchronised, or at least market efficiency; and if it declines it might indicate non-economic sustainability (see Figure 6.1).[2] This tells us already at a citywide level about the sustainability, but subsequently the same needs to be carried out for city districts (and house types) in order to examine how the trend varies by district location (and house type). The further aim then is to see whether a positive economic development, as

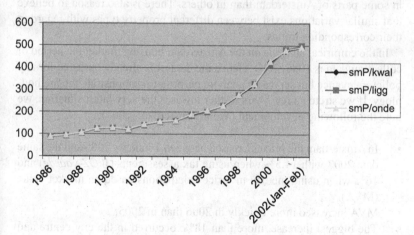

Figure 6.1 Time-series of the square metre price related to three different quality measures aggregated for Amsterdam.

measured through increasing house prices, leads to spatial (or functional) differentiation like general spatial economic theory predicts.[3] The square metre transaction price (smP) represents an economic criterion, whereas the quality of the house (kwal), the level of maintenance (onde) and the quality of the location (ligg), respectively, refer to the three different physical quality criteria: dwelling quality, maintenance and quality of micro-location, respectively. Figure 6.1 shows how the steepness of the trend increases around the year 1998.[4]

Figure 6.1 shows clearly the measure of inflation in the Netherlands, which during the time-period of collecting data was substantial. However, differences between regions in the Netherlands do exist in this respect. Here it has to be noted that, compared to the rest of the country, the Randstad region (the central-western part of the country) including Amsterdam is economically unsustainable in the sense that, in average terms, the dwelling price is inflated in relation to affordability (Op't Veld et al., 2008). This region has also experienced a stronger increase than elsewhere. (This was also noted in Chapter 2 in this volume.) There are similar but lesser differences between districts within Amsterdam, in particular the districts in the Western part have not experienced as steep increase than the rest of the city (see Figure 4.8).

Four disaggregated trends are shown in Figures 6.2 and 6.3: two of them (the upper graphs) have near identical magnitudes with the aggregate trend and even a steeper price increase after 1998, whereas the other two (the lower graphs) show a considerably less steep increase in the price-to-quality ratios. Thus the price development 1986–2002 was more sustainable in some parts of Amsterdam than in others. There is also reason to believe that similar variations exist between different property types with regard to their corresponding trends.

In the empirical analysis on the Amsterdam housing market the period of collecting data is 1986–2002. Hence the dataset is already outdated for any policy relevance but serves an analytic purpose to demonstrate the methodology. If we stretch a few years further towards the very latest situation, we note the following development (Woon Amsterdam, 2007):

- In Amsterdam the price increased between 1 January 2005 and the same date 2007 with 14.5% when using tax assessments (*WOZ-vaarde*) and 16% when using median m² price calculations based on market value (MVA).[5]
- MVA increased more quickly in 2006 than in 2005.
- The biggest increase, more than 18%, occurred in the city centre and the south – the most popular area; the smallest increase, less than 9%, took place in the north and west (see Figure 4.8).

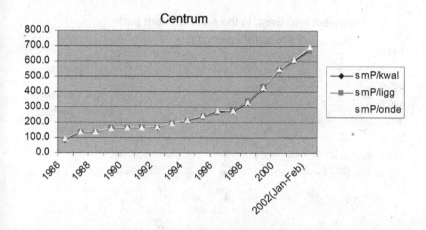

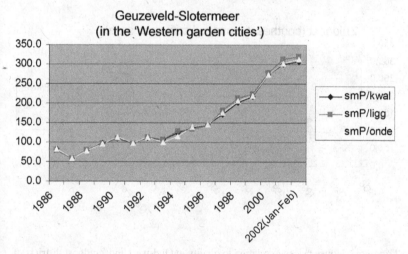

Figure 6.2 House prices in relation to quality in Centrum and Geuzeveld-Slotermeer districts of Amsterdam.

- The Amsterdam housing market is very tight. As a consequence the offer-prices are getting closer to sales prices and are sometimes even below them!
- The most popular type of apartment in terms of quickest price increase varies across districts (using four categories): in the city centre it is the pre-war type; in some other districts it is those apartments built 1945–1970, or even after 1970; yet elsewhere it may be the single-family homes.

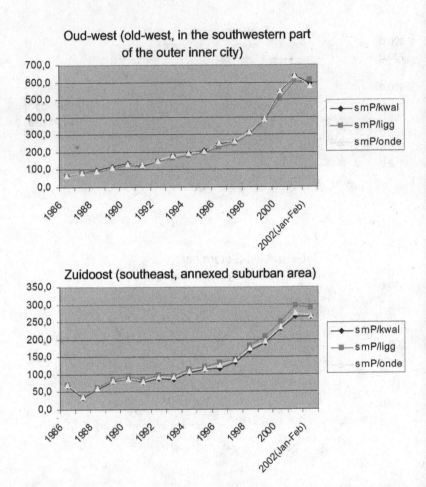

Figure 6.3 House prices in relation to quality in Oud-west and Zuidoost districts of Amsterdam.

Thus, while we deal with a chronically unsustainable market, we also deal with great variation across locations and house types and with a remarkable time-variation.

LVQ classification

The next procedure is to generate SOM outputs for each cross section for Amsterdam. We also need to calibrate each SOM output with the P/Q indicator as label. Here the three citywide aggregate P/Q indicators for quality are reduced into one (simply by taking the mean); then this indicator will be rounded into three ordinal categories for labelling. Finally, we apply the LVQ classification using this labelling as classification criterion. In principle, we can also compare this with other classification criteria, but the main thing here is that we compare across the one-year cross sections. It is to note that only the LVQ results (not the SOM results) are reported here. For the labelling with the best classification result the distinction between sustainable and unsustainable locations must then be assumed to be important. We define the following based on earlier theoretical propositions (see Kauko, 2008a). If the P/Q indicator is below a lower pre-specified threshold, the case is sustainable in non-economic terms; if this indicator is above the lower threshold but below an upper pre-specified threshold, the case is economically sustainable; and if this indicator is above the upper threshold it is totally unsustainable. A label will be assigned *ad hoc* as follows: 0 for non-economic sustainability (P < Q); 1 for economic sustainability (P = Q); 2 for total unsustainability (P > Q). When classification runs are carried out systematically for each cross section, we get a formal measure of the strength of discrimination of each of the three classes as well as an average measure of the whole cross section. We furthermore obtain a time-trend as to which of the three classes is increasing or decreasing its influence as discriminator of the classification as well as how the average classification for the whole sample develops through time. Figures 6.2 and 6.3 each show that the upper graphs have an index of more than double that of the lower, which means – given our definitions – that only the lower market development is sustainable in relative terms (i.e. at least more economically sustainable than the upper situation). It is, however, to observe that this diagrammatic illustration concerns only the economic dimension(s). In fact, the Geuzeveld-Slotermeer district – a typical Dutch unfashionable garden city milieu – has lots of social problems. In Figures 6.2 and 6.3 two turning points are to note in the lower graphs: around 1991 and 2001 respectively, when the P/Q ratio 'dips', in other words, turns more sustainable.

The method in the Amsterdam analysis of P/Q ratios can be summarised as follows:

- We define the following three classes of sustainability:

  - If the P/Q indicator is below a lower pre-specified threshold, the case is sustainable in non-economic terms (labelled 1).
  - If this indicator is above the lower threshold but below an upper pre-specified threshold, the case is economically sustainable (labelled 2).

   • If this indicator is above the upper threshold it is totally unsustainable (labelled 3).
• When LVQ classification runs are carried out systematically for each cross section (during the period 1986–2002), we get the following:

   • Comparison of the strength of discrimination for each of the three classes and an average for the whole cross section.
   • A time-trend as to which of the three classes is increasing or decreasing its influence as well as how the average classification for the whole sample develops through time.

The resulting percentage figures are shown in Table 6.2. What do these figures tell us? When we look at the category specific figures, 1993 was a 'totally unsustainable' year, 1996 in turn an 'economically sustainable' year and all other years 'non-economically sustainable'. This means that, apart from a few exceptions, the outcome over the whole period is clearly on the sustainable side. As for the overall figure, the interpretation is that at first the trend in recognition accuracy is a falling one, then stable with small fluctuations and finally increasing. This indicates a growing ability of the data to discriminate between the three defined cases in relation to sustainability,

Table 6.2 The result of the LVQ classification.

Year	N (total/test)	LVQ-accuracy for the whole data	The same % for each category		
			0 (non-economic sustainability)	1 (economic sustainability)	2 (total unsustainability)
1986	480/48	95.83	100	0	–
1987	579/58	65.52	100	0	0
1988	692/69	57.97	100	0	0
1989	827/83	72.29	100	0	0
1990	1,017/102	50.98	100	0	0
1991	2,642/264	66.29	95.11	0	0
1992	2,680/268	52.61	96.58	0	0
1993	2,925/293	1.71	0	0	100
1994	3,638/364	65.38	98.76	0	0
1995	3,313/331	49.85	100	0	0
1996	1,891/189	47.09	0	100	0
1997	3,345/335	68.66	99.57	0	0
1998	2,930/293	74.74	100	0	0
1999	3,557/356	57.87	100	0	0
2000	4,053/405	76.30	100	0	0
2001	3,799/380	81.32	100	0	0
2002	70/	Too few observations to perform any meaningful analysis.			

after an initial period of decreasing ability to discriminate. In other words, the class boundaries between the two cases were first sharp, then became less sharp and finally sharper again. This indicates a growing difference between the sustainable majority and the unsustainable minority cases in the dataset throughout the sixteen-year period.

While only providing an indication of the tendencies goings on, these results seem promising. Here is plenty of scope for further work in relation to economic and other sustainability classification/analysis of different areas, property types and particular indicators/criteria applied. The obvious caveat here concerns validity of the analysis as the method is only in a testing stage. To what extent can we consider the dataset for one-year transactions large enough? We note that the last year is not sufficiently covered by data to be able to generate a meaningful classification ratio.

This chapter has demonstrated another empirical application of the idea of socio-economic sustainability metrics in relation to house price data. It follows the methodological approach applied in Chapter 5 and, on a more general level, the overall argumentation in Chapters 1–4. Sustainability analysis is a multifaceted topic and the line pursued so far is focusing on a few key issues: social and economic dimensions, spatial elements and testing of alternative methods. In the next chapter yet one more empirical analysis is documented in this vein. (This comprises the penultimate essay of this compilation.)

Notes

1 In the categorisations, the size of the house and plot are split into four areas: A, B, C and D. A represents the main part of the area that is in use, while B, C and D are negligible areas such as cellars, sheds or lofts. In the selected definition of the variables 'transaction price per m²', 'house size' and 'lot size', only the area A is included for each transaction.
2 The house prices recorded are nominal, but the real prices could be obtained by deflating with the inflation rate 1986–2002. That would be a figure in the interval of 1–4. The threshold values for the three sustainability categories were deflated with an annual rate of 3%.
3 Unless the starting point is low price level, in which case any price increase would lead to a homogenisation (e.g. Bramley et al., 2008, p. 194).
4 Without further elaboration (as this issue goes beyond the purposes of this book) we note that this is incidentally also the year when a market deregulation and privatisation begun in the Dutch housing policy.
5 Here it is to note that the *WOZ-vaarde* includes estimates for rental apartments. As these have on average lesser quality than owner-occupied apartments, it is logical to find the increase in *WOZ-vaarde* somewhat lower than that based on market values only.w

7 Sustainability of urban renewal areas

Cases of Budapest and Amsterdam[1]

Debates on sustainability in an urban real estate context share much of the debates around the general concept of sustainability. To reiterate the overall aim of the book, sustainability – in the built environment context – is a multifaceted concept: there is much more to it than the 'green' issues usually on the agenda. Why the less explored aspects of sustainability are not equally noted has to do with the yet emerging discourse on exact definitions and, the lack of credibility when dealing with socio-economic and, perhaps more so, with socio-cultural sustainability issues within urban real estate. To make analyses of these aspects credible, the argument here is that a metric is necessary. Such a metric, for example based on property prices, undoubtedly would give an added value for real estate analysis. In the previous two chapters modelling and classification was carried out. In the present chapter a simpler approach is applied: to split the data of residential property transactions in a neighbourhood into target and comparable groups. The specific topic here is urban regenerations, which is why the sample is selected in such a way that the cases are either assumed affected or unaffected by a given planning or policy measure.

When examining urban property development from an economic geography perspective the issue is about producing and maintaining an advanced producer service (see Bryson, 1997; Hamnett, 2003). On the other hand, real estate occupies a special position among industries, which causes two kinds of problems for the analyst. First, the difficulty of merging residential and commercial property: the former applies primarily a microeconomic approach and involves welfare economic elements (housing policy), whereas the latter applies a macroeconomic approach involved with business economic considerations. Second, another problem caused by the spatial fixity of the product which precludes abstraction in the sense favoured by current social and spatial theorists. Because of such difficulties real estate is often left out of the mainstream research topics within the field of urban research.[2]

In the present contribution, urban regeneration can be defined as the management and planning of the existing urban tissue as opposed to new areas (Couch and Fraser, 2003). This chapter presents an analytical framework for drivers, motivations, trajectories and sustainability of locations subject to urban regeneration. The following *research questions* are posed in a comparative perspective:

- What are the motives of actors?
- What kind of private-public balance prevails?
- What drives the renewal and refurbishment processes?
- Are these drivers primary related to economic structural conditions, or to the system of subsidies and tax incentives or to the nature of the motivations for certain types of urban regeneration projects?
- And within the latter category: is it primarily about profitability or is it about sustainability – this dichotomy is referred to as 'the double bottom-line' (see Bryson and Lombardi, 2009)?
- Further to this, does the development strategy comprise demolishing and new development of blocks, or is a lighter and cheaper strategy opted for instead, that is to say refurbishment of the dwellings, communal area or façades?
- Last, what about the people living in the regenerated neighbourhoods – do they have a substantial role beyond merely being faceless market actors?

This problem area is highlighted using case studies from Budapest and Amsterdam. The method is simple: comparison of homes sold when these are assigned to two groups: (1) subject to regeneration, and (2) not subject to regeneration, but situated nearby. The results show that Amsterdam is more sustainable than Budapest for both the supply side and demand side processes of property development. This side of the sustainability issue has to do with the balance of social vs. private housing: the more there is social housing in relative terms, the more the use value matters. Only profit characterised the Budapest cases, whereas the use value too was considered in the Amsterdam cases. Moreover, what is interesting is that in both cases *a shift towards lesser government control* has taken place in recent years.

Integrating explanations of the initiation of urban change

Rent gap concerns the difference between potential value and current value. Following Clark (1987), the value gap composes a part of the rent gap: that between capitalised contract rent and the market-based sales price of the dwelling. Closing the value gap then implies partial closing of the rent gap. The complete closure of the rent gap nevertheless implies realising the potential rent or price level of the dwelling. This understanding was for a long time compatible with other views of urban change. However, since the initial consensus the significance of rent gap and even gentrification has been played down; this explanation of gentrification is supply sided and involves only economic value and profitability calculations, although the value gap theory also pays attention to rent and price regulation aspects (see also Jäger, 2003).

Zietz and Sirmans (2004) review the literature on inner-city property market analysis and conclude that it is likely that reaping positive returns from revitalisation efforts requires government intervention. They emphasise the investment perspective and the need for sound investment and policy strategies pertaining to the inner-city environment. Zietz and Sirmans furthermore note that the redevelopment of the inner city involves several spatial issues, gentrification being one of them, albeit an overvalued target for research. It is also not clear that the rent gap always is connected to gentrification, according to these authors.

Whether the driver of urban regeneration then is more *demand* or *supply oriented* is an empirical issue subject to scrutiny in the empirical part of the study. Another empirical issue concerns the balance between two different development strategies: 'exchange value only' (profit-making) and 'use values too included' (i.e. non-economic sustainability). At this point it is sometimes argued, incorrectly, that suppliers pursuing profit maximisation in relation to the building process do not ignore use values, because the demanders of space incorporate use values into their price bid, either directly (owner-occupiers) or indirectly (investors' interpretation of tenant-occupiers' requirements). According to this logic then suppliers' profits will be affected negatively, if this is not taken into account. This is, however, a flawed conclusion: it makes sense to assume that profit maximising builders and developers can cause artificial scarcity and thus a situation in which anything goes: even 'chicken boxes' without any use values whatsoever. Thus, in such a situation there is 'nothing new under the sun' as the non-market use values are indeed ignored, and thereby missing the opportunity of creating more sustainable built environments.

From active to passive government involvement

Government involvement in local housing markets is a particularly interesting arena for this study. According to Hamnett (2003), while we to some extent can speak about general processes, different cities respond in different ways to the dynamics of globalisation (see also Kovács, 2009). Why is a particular urban renewal measure implemented in area A, where the role of government is active – 'making it happen'? Why is gentrification taking place in area B, where the role of the government is passive – 'letting it happen', that is to say, weak local social policies combined with favourable incentives that are offered for landlords (e.g. housing associations) to sell their stock and for owner-occupiers to buy dwellings. Thus two opposite situations ought to be taken as starting points for the conceptualisation. Loosely following Hamnett (2003) a distinction may be made between the two ideal types of government roles:

(A) Actively pursue an idea of regeneration on either grounds:

- Social-economic (and political)
- Physical-functional (as will be shown in the case study of de Pijp to follow).

Policies such as New Urbanism and Neo-traditionalism fall within this category.

(B) Passively sitting back and hoping that there will be organic gentrification, utilising the 'neo-liberal' market stimulation policies of the state, when the following preconditions are favourable:

- Incentives for landlords to sell in order to gain profit (as a result of price inflation),
- Incentives for individuals to buy in order to gain favourable mortgage finance,
- Weak protection of tenants and use value considerations (as a result of state level deregulation of particular housing and planning policies).

When we use Hamnett's categorisation between more active and passive government measures we can find out tendencies for either government initiated urban regeneration or market driven, organic gentrification. Some cases are, however, of borderline nature and thereby difficult to classify (see Figure 7.1). For example, the London Docklands Development Corporation was appointed by the State (thus, not the local) government – this is nevertheless a variant of the active government type. Uitermark and colleagues (2007) in turn identified 'state induced gentrification' in Dutch cities – this is probably closer to the active type as the state actively promotes the area for potential gentrifiers.

Beginning the interpretation of the figure in the upper right corner, the global cities such as Amsterdam and Budapest largely attempt to attract the creative knowledge professionals of the New Economy or cultural industries. Here the government role might be passive and leave the development

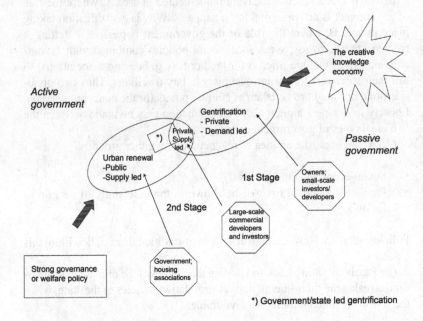

Figure 7.1 Schematic illustration of the forces at play in urban regeneration and property development.

to the market forces. As these groups of people also need accommodation close to (or within good access to) their jobs, the gentrification process of relevant neighbourhoods begins. In the first wave, the owners themselves renovate their dwelling and then sell them to these new occupants or to small-scale investors or developers. Of course, it may be that the owners are able to capitalise on the increased demand for home buying even when no renovation has taken place. In the second wave, housing associations sell their rental flats with similar motivations. As the bulk of flats enter the market, the buyers might now be large-scale developers and investors too. In the third wave we enter a situation that is 'breakwater' between where gentrification ends and urban regeneration begins.

The exact point of difference between gentrification and urban regeneration is impossible to determine, however. The urban regeneration process has another motor than traditional gentrification has, namely the active government, who attempts to revitalise derelict neighbourhoods – or whole disadvantaged cities and city-regions. It may also be that this role falls to another actor such as the state or private capital. Often (but not always) PPP becomes the most effective means to initiate and steer this process. On the other hand, to determine the ideal PPP structure for a given project is notoriously difficult (see e.g. Taşan-Kok, 2010).

Various preconditions and forces mould the patterns of urban regeneration, and this has further implications for the housing market of the affected neighbourhood. Whether the local government sits through the process as a predominantly active or passive player, whether the supply or demand side is the driver, whether the development is small scale or large scale and whether the primus motor is public or private sector investment, are all issues that will affect the outcome in spatial, economic, social, aesthetic and environmental terms. There are other issues too such as technology, legal frameworks and customs in initiating and directing these trajectories. There is furthermore risk in being 'creative' in this context. Obviously, both kinds of extremes (i.e. totally active or passive government) are subject to criticism here.

The discussion so far has concerned both market level processes and government responses to them. This dialectic is important for the remainder of the study, given that how public sector directs the urban development forms an important backdrop for the analysis of urban property development (see also Adams et al., 2005). The case studies documented next are from four neighbourhoods in two European capital cities: Budapest and Amsterdam.[3]

Case studies of Budapest

Tsenkova (2006, pp. 351–352) notes that increasing polarisation and deprivation is common for inner cities and peripheral housing estates in post-socialist cities and that this leads to a pattern of poverty concentration in run-down neighbourhoods and to increasing costs of urban services and housing. Indeed inner Budapest is much shaped by pre-socialist times, and then neglected by more than forty years of socialism – also socially. Before 1990 rehabilitation plans were implemented only in few sites in the inner city. After 1990s municipalities sold their shares to the new owners, which led to new development trends in the urban space. For sustainable urban development and renewal some of these trends are favourable and others are not. On the other hand, as the slumming of the inner parts of the city was causing damage also in terms of the viability of the whole city, urban renewal was considered a possible alternative for greenfield development and abundant suburban development (Locsmándi, 2006).

It is to note that during the first half of the 1990s the private sector had little interest in the renewal of residential buildings, apart from a few exceptions. This changed in 1994 when the blocks of flats obtained a firm legal status in the Act on Condominiums. Then in 1996 the City of Budapest developed an official urban rehabilitation programme. The regulations of this programme stipulated that housing condominiums located in those districts that had transferred 50% of the revenues of the privatisation of public rental dwellings to the City government budget could apply for support from the Rehabilitation Fund (Kovács, 2009).

In Budapest a system of subsidies for urban renewal was available for both local government and household group initiatives. Here the actors were expected to actively acquire government subsidies from the Rehabilitation Fund. A great part of these subsidies was awarded to district IX. In district IX the local government decided only once per year which buildings were to be sold; otherwise they did not want to be involved in the process, as the private sector was responsible for the building. However, the district used the income obtained from transferring the municipally owned land into a private-public development company to improve public spaces and infrastructure.

At the time, in Budapest the supply side driver of urban regeneration clearly was the profitability of the housing development or urban regeneration project. Also a demand-driven gentrification may be true in some pockets in Budapest, but in general the issue is not the same as in cities in the UK (cf. Cameron, 2006), or in de Pijp – one of the neighbourhood cases of Amsterdam chosen for this study (see next section). The demand side driver, tenure change initiated/triggered by tax benefits or mortgage financing is not an issue in Budapest; neither is the supply side goal of social or physical comprehensiveness such an issue of relevance as the developer has a goal in improving only the 'market' and the 'value' – not the 'social' side of it.

It is often asserted that privatisation of dwellings is counterproductive to comprehensive renewal. Földi (2006, pp. 118–125) concludes that functional conversion from apartment to offices and from lofts to apartments took place elsewhere in the inner city except in the districts VIII and IX. In these two districts privatisation was blocked, which resulted in substantially more preserved old housing units than elsewhere. In fact, in the other Pest side areas cut by the Grand Boulevard (i.e., the districts VI and VII) apartments were sold mainly for offices and hotels. In the most dilapidated part of district VIII, where the share of public rental stock remained as high as 25%, the higher share of public housing facilitated a gradual social renewal. Another economic development is that the inhabitants of condos increasingly formed solidarity groups commonly interested at least in partial renewal of their buildings (Locsmándi, 2006).

According to Földi (2006) profit-making is the key to analyse the upgrading of residential status of Budapest neighbourhoods (cf. Cameron, 2006: in UK). She notes that a massive rent gap occurred in some neighbourhoods during the 1990s, and as a consequence, the market has shaped these areas – often in a two-stage process: (1) speculation stage, and (2) building stage. She notes that 'status upgrading' took place in district IX at the price of neglect of community building. When the social aspect is weak, the result is a replacement of the old population with an upwardly mobile new population.

In Budapest during 1995–2010 hardly any comprehensive urban policy making related to housing or real estate took place. In attractive areas the market was expected to take care of the development. As it was furthermore accepted that in other areas the passive planning system would be unable to improve the situation, these areas were left derelict. Here any active planning lacked resources and political support. This was much related to who was in power and where. Notably, after the year 2000 the Hungarian right-centre coalition government launched a programme for the construction of new public rentals, but its impact was negligible in Budapest where the liberals and socialists had been dominating (until the local elections of year 2010). Földi (2006) asserts that differences in district administration influence the urban renewal strategy, which is a unique feature in CEE. She concludes that the neighbourhood dynamics in the Budapest inner city varies across and within districts in relation to the type of renewal process.

In the district IX a differentiation has occurred since the early/mid-1990s. The current situation is that the area adjacent to the city centre (Kalvin sq., Raday st.) is completed, in the middle part (between the Grand Boulevard and Haller street) lots of demolishment and indeed revitalisation has occurred, and that the outermost part of this district (the railway bridge and beyond) is a good example of intended brownfield development. In fact, along Soroksári ut a massive development (Duna City) is planned. This project is rather ambitious, as it would mean a moving of the city centre *c.* 1 km southwards! However, this project is still short of investment (even at the time of writing this, in spring 2013), and due to the current financial crisis its continuation is unsure.

The research design of the current study follows traditions of 'quasi-controlled' experiment and institutional analysis. In the Budapest cases the urban renewal and rehabilitation outcomes were compared partly based on house price trends at the street level (dataset of the Hungarian statistical office, KSH) and partly based on narratives and documents such as interviews of stakeholders and experts and official accounts on the development plan (IX) and development strategy (VIII) as well as marketing documents (brochures). When relevant, the street was disaggregated by the two main house types: condominium or panel houses. The areas are shown in Figure 7.2.

The district VIII comprises totally different areas: *the inner part* (Palace quarters, the northeastern portion of the map and marked D in Figure 7.2),

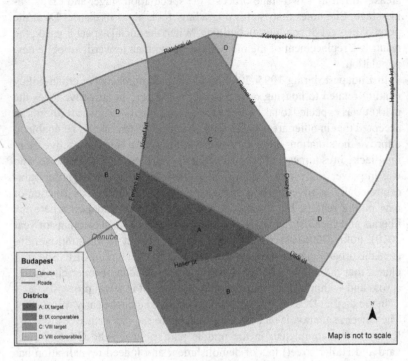

Figure 7.2 The case study areas of Budapest: target and comparable areas.

which also is the densest area, is without doubt the most attractive neighbourhood; the Grand Boulevard is the cutting line between the inner and
middle parts; the outer parts (beyond Kerepesi cemetery, the white zone)
are far away and represent different area types altogether. The target area of
the district IX (A) is characterised by mixed land use and piecemeal, small-
scale development much in accordance with the principles of Neo-traditionalism. This area has become more diversified and heterogeneous in terms
of the price levels and housing types than the neighbouring renewal area in
district VIII (C), where a lot of the stigma apparently has disappeared, and
where prices have increased everywhere. This latter area is, however, still
evolving and one cannot draw definite conclusions – *Corvin Promenade*,
the biggest project of this district, is in fact being developed according to
modernist design principles involving large-scale new developments.[4]

On a general level, the street and district-wise aggregated dataset (KSH)
tells us that, particularly for condominiums, prices in district IX by far
exceed those in district VIII. Clearly this house type, which composes the

majority of the dwelling stock in the Budapest inner city, is considered a more attractive choice in district IX than in district VIII (see Figures 7.3 and 7.4). The exception being the brand new area of condominium blocks constructed in the southwest corner of the C-area shown in Figure 7.2: this is *Corvin Promenade*, a large-scale project comprising a shopping mall, modern offices and homes with the character of modest 'residential parks' (see Figure 7.5). According to the results the nature and pace of the changes are different in the two affected target areas, when related to unaffected comparable areas just outside: in district IX the price level of the target area is higher and the increase less steep than in the comparable area, whereas

Figure 7.3 Streetscape of middle district IX.

Figure 7.4 Streetscape of middle district VIII.

in district VIII the price level of the target area is lower and the increase steeper than in the comparable area (see Table 7.1).

Bramley et al. (2008, p. 194) claim that an increased price level in general brings a lesser variation in the housing packages in a given neighbourhood, this being based on evidence from neighbourhood housing markets in England. However, based on the Budapest analysis an additional qualification is needed: such a development would apply only for neighbourhoods with

Figure 7.5 Corvin Promenade (middle district VIII).

Table 7.1 Mean price development (1000 HUF) for *condominiums* disaggregated by area (target or comparable) and citywide.

	1997	1998	1999	2000	2001	2002	total change %
IX target	65.2	70.0	78.5	117.4	145.2	175.5	+169
IX comparable	58.2	62.1	72.1	106.9	132.1	159.5	+174
VIII target	40.8	44.7	53.0	92.6	107.0	124.4	+205
VIII comparable	49.5	55.3	64.1	108.7	119.5	133.5	+170
Citywide	63.0	67.0	77.0	129.0	162.0	170.0	+170

Note. Calculations by the author using data from the Hungarian Statistical Office (KSH), aggregated per street and district (municipality).

low starting point in prices – already the average price level of the neighbourhood in the middle part of district IX shows the opposite trajectory. Thus, when an originally average level area such as the middle part of the district IX is being upgraded, the socio-economic and physical features as well as prices in fact become more differentiated (and the area structure on balance more heterogeneous) than before due to the increased significance of consumer preferences. This broadened variation in itself indicates economic sustainability; as conceptualised in the previous chapters, diversity is a favourable development from a sustainability point of view. Furthermore, when an originally low-level area such as the middle part of the district VIII is being upgraded the socio-economic and physical features are becoming more equal (and the area structure on balance more homogeneous) than before due to the gradual disappearance of concentrations of unfavourable and derelict housing.

Case studies of Amsterdam

In the Netherlands traditionally a strong planning system prevails. The Dutch planning tradition is according to Kloosterman and Lambregts (2007) 'rooted in an outspoken modernist and static view of the world'. A new plan tends to be designed approximately every ten years, and this typically incorporates some aspects of urban renewal too. On a general level, however, the conditions for a regeneration process are shifting away from a more active local government towards a more passive government and this applies for Amsterdam, the capital city, too. Nonetheless, the urban regeneration process has retained its complexity, as it has a number of particular supply and demand side drivers (i.e. production and consumption motives). On the production side also sustainability goals are included. On the consumption side, the renewal is, however, not so much motivated by tenure change as the majority of the stock in this city remains owned by housing associations.

Amsterdam is sometimes mentioned as an example of successful urban regeneration, as in this city regeneration has resulted in more affordable housing, a reduction in carbon emissions, more efficient recycling rates and increased community cohesion (RICS, 2008). At the same time, however, interethnic cleavages in Amsterdam have become accentuated in recent times, given that the share of Natives in the city as a whole is only 51.5% (in year 2006), the Turks (5.2% share) being the most segregated ethnic group (Smets and Kreuk, 2008). In order to combat such socially unsustainable situations, the City of Amsterdam is able to build owner-occupied housing also in less favoured areas, because it owns the majority of the land.[5] For example, in de Pijp – the other case study area – buildings used to be among the worst quality in the city, but later the allocation of the most expensive housing stock took place here. Moreover, Amsterdam suffers from a chronically tense housing market, where status, social externalities and the sense of belonging define the attractiveness of a neighbourhood rather than the physical environment.

In order to widen the context of the multiple case study – and thereby the generalisability – a pair of case studies of neighbourhoods was carried out in Amsterdam. Two cases were selected: in the western (Geuzenveld-Slotermeer) and southern (Oud-zuid) sectors of the city, respectively (see

Figure 7.6). The two neighbourhoods and districts are rather different and situated far away from each other.

For *Buurt Negen* urban renewal and redevelopment projects are meeting the demand for a higher level of single-family homes and owner-occupied (higher quality) dwellings. The ethnic aspect is particularly notable in this part of the city (Smets and Kreuk, 2008). The first concrete renewal processes in Buurt Negen began 1997, partly on brownfield, partly on greenfield sites as a PPP. However, during the implementation phase problems occurred, mainly due to the lack of coordination between district and city departments. One of the bottlenecks encountered within the urban restructuring process was the different nature of the projects of housing associations.

In *de Pijp*, a neighbourhood in the southern sector within the inner suburbs (see Figure 7.6), the pro-market change in policy has led to an upgrading. In the beginning of the 1970s, the share of public investment was higher (and share of private investment lower) than today. While the first new dwellings were built in 1960s, plenty of renewal took place in the 1980s, including concentrations of newly built blocks involving functional change. The main actors were the housing corporations on one hand, and on the other individuals, including renters, homeowners themselves, and investors, who own a small number of dwellings, and since the late 1990s,

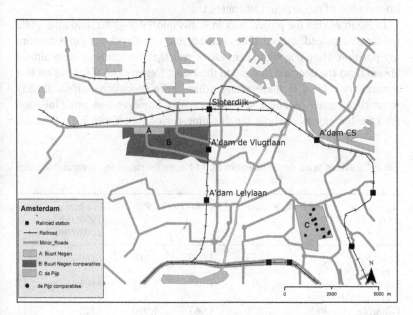

Figure 7.6 The map of case study areas in Amsterdam: target and comparable cases.

a new corporation, *Ymere*, as a successor to the City ownership. Together with the abolishment of subsidies noted earlier, this devolution of the city's housing stock into a new corporation represents a major shift towards market led housing policy, and it is likely that this shift has had a major impact on subsequent revitalisation processes too.

In the present study the urban renewal and rehabilitation outcomes were evaluated partly based on house price trends at the individual transaction level and partly based on narratives and documents such as interviews of stakeholders and experts and official accounts on the project implementation (Buurt Negen) and the development plan (de Pijp). Like with the Budapest cases, quantitative data was linked with case study material and statistics on house prices. The idea was to compare the price development of certain identifiable micro locations inside and outside the target areas and see what the connections are between the general price development trend on one hand and the building up of the area on the other.

A micro-level dataset used for taxation purposes in Amsterdam consists of more variables and observations than in the corresponding Budapest dataset, and it is also more reliable, where indicators of maintenance, dwelling quality and micro-location quality were utilised and related to the price changes in time for particular neighbourhoods subject to revitalisation. (A dataset comprising the sales prices of *c.* 46,000 dwellings with recorded information of the selected attributes.)

In Buurt Negen the project was in status quo 1998–2002, and after 2002 the terms changed: the subsidies ceased, and one developer got a monopoly position. Here one significant factor was that the residents were almost absent from the decision making in the Buurt Negen case. It is obvious that uncertainty related to these changes has had a speculative effect already some time earlier on house sales. There was, however, less price inflation than for the city as a whole for this time-period (see Table 7.2). Thus the

Table 7.2 Mean price development in Buurt Negen for target and comparable cases and citywide* for *row houses.*

Sales price NLG	1986–1989	1990–1993	1994–1997	1998–2001	% change 1990/1993– 1998/2001	N
Buurt 9 – target		1,121	1,069	2,116	+89	5
Buurt 9 – comparables		913	1,047	1,832	+101	57
Citywide	963	896	1,063	1,990	+122	1,444

*Selected from the whole Amsterdam dataset; no suspiciously low price, not canal situation.
Note. Calculations by author using data from the City of Amsterdam Tax office.

Table 7.3 Price development in de Pijp for a block with three or more floors and built 1915–1945.

Sales price NLG	1986–1989	1990–1993	1994–1997	1998–2001	% change 1986/1989– 1998/2001	N
de Pijp – target	534	648	1,296	2,255	+322	34
de Pijp – comparables	555	770	1,392	2,111	+280	15
Citywide	629	1,120	1,383	2,665	+324	6,139

Note. Calculations by author using data from the City of Amsterdam Tax office.

price increase was partially moderated by components with a spatial dimension that affected only the part of the city under study.

In de Pijp it was possible to identify ten blocks without renewal, where nothing had taken place and also nothing had taken place opposite or next door. The Table 7.3 shows the corresponding results for *de Pijp*, where the target cases situated in the interwar segment has been subject to a price lift during the time-period under study.

The analysis of how the quality developed for this house type in de Pijp is shown in 7. The only increase in the quality-rating concerns the maintenance indicator. This analysis is, however, not as informative as the corresponding analysis of price development. Namely, in order to make the data comparable, the data were first screened based on the criteria of being of not too dissimilar quality. More specifically, on a scale of 1–10, commonly used by valuers commissioned by the City of Amsterdam tax office (Gemeentebelastingen), all the selected cases have an assessed mark within the range 6–8 all for three quality attributes: quality of dwelling, maintenance and quality of micro-location. Nevertheless, Figure 7.7 a–c suggests that, for two of the three recorded quality attributes, the quality of the target cases did not rise during this time-period, despite casual observations and media coverage telling the opposite! On the basis of this evidence it would therefore be too speculative to conclude that any tangible improvements in the quality of either the dwellings or locations has taken place in the target area. These findings do not therefore suggest that any substantial quality improvements, ostensibly taking place as a result of increased demand sided market influence (i.e. the shift from active to passive government role), have caused the observed price increases in de Pijp (see also Kauko, 2009a).

While detailed and more comprehensive data make conclusions less black-and-white than in Budapest, we can still cautiously conclude that no price increase for the target cases took place in most of the Amsterdam areas/segments under study when compared to non-affected cases and citywide figures. Of the two cases, only in de Pijp a sharp price increase was

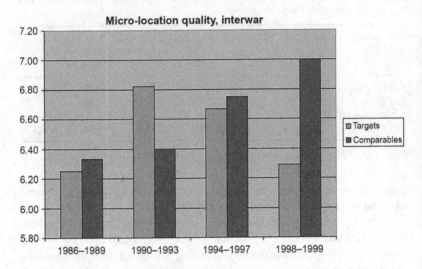

Figure 7.7.a Quality development in de Pijp for block with three or more floors and built 1915–1945: the quality of the micro-location.

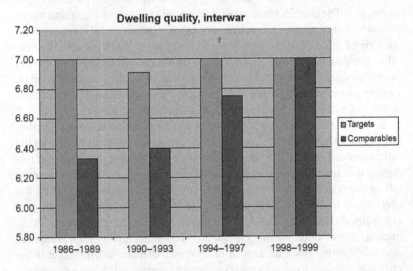

Figure 7.7.b Quality development in de Pijp for block with three or more floors and built 1915–1945: the quality of the dwelling.

identified, but it is also true that in this area plenty of improvement has taken place, which would indicate that any localised price hikes are the consequence of the tangible, micro-level improvement of the amenities or just improvement of the streetscape and façades.

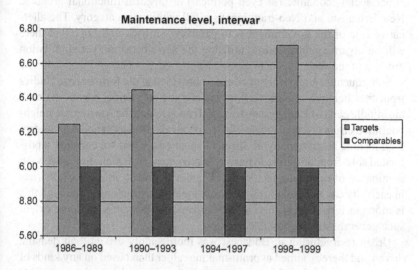

Figure 7.7.c Quality development in de Pijp for block with three or more floors and built 1915–1945: the maintenance.

Summary of the case studies

According to Hamnett (2003), while we to some extent can speak about general processes, different cities respond in different ways to the dynamics of globalisation, depending on the existing economic structure, level of development, history, position in the global economic and financial system, social and ethnic composition and the nature of specific policies in relation to immigration, labour market, wages, housing and planning. The study documented in this chapter pays attention to the last two of these factors. Why was a particular urban renewal measure implemented in area A, where the role of government was 'to make it happen' (active)? Why was gentrification taking place in area B, where the role of the government was 'to let it happen' (passive), that is to say, weak local policies and favourable incentives were being offered for landlords to sell their stock and for prospective owner-occupiers to buy their dwellings?

The four case study areas were examined with respect to price development, quality development including an intangible element (image) and whether the role of the government was active or passive in their restructuring strategies. This last factor was particularly interesting for this study: exactly which measures generate urban generation or exactly how does gentrification occur. A distinction was made between whether the role of the government is 'making it happen' or 'letting it happen'. The former situation arises when the government actively pursues an idea of regeneration on either social-economic (or even political) or physical-functional grounds. New Urbanism and Neo-traditionalism fall within this category. The alternative role of the government is to passively sit back and hope that there will be organic gentrification, utilising the neo-liberal market stimulation policies of the state.

Subsequently, empirical material was provided in the form of case studies from four neighbourhoods that are undergoing substantial rehabilitation – the middle parts of Budapest districts VIII and IX, and the Amsterdam neighbourhoods Buurt Negen and de Pijp, respectively. Budapest and Amsterdam both include heterogeneity of different situations, when we consider a postulated shift from an active to passive government type including a growing dominance of economic factors in urban development. In fact, the districts in each city can autonomously carry out their policy making – although this is more so in Budapest than in Amsterdam. Nevertheless, to some extent such generalisation is possible here too.

Urban restructuring in Budapest was more supply driven than demand driven and thereby aimed at profit-making rather than based on any kinds of rental or use value considerations. As a consequence, this pushed up prices 'artificially', for example, through the large-scale expropriations carried out in district VIII by its property management authorities. In Amsterdam the observed change was demand driven. More specifically, the driver was tenure change from rental to ownership in order to capitalise on the exchange value (e.g. due to tax benefits) or because of use value considerations of the house-buyer. Thus in some circumstances the demand side motives generate an increase in prices, and in other circumstances the price lift is caused by supply side motives. Both kinds of processes can lead to a situation where the price increase is unrelated to a corresponding quality improvement. In such a case the price is also not in relation to the capitalised rental income anymore, which per definition indicates a disequilibrium mechanism, the empirical manifestation of which verifies the rent gap (or value gap) hypothesis.

In Budapest the general strategy comprised block-wise and developer-led market rehabilitation of the housing stock, using a system of subsidies available for both local government and household group initiatives. Here, however, the revitalisation and development processes differed across districts. In district IX they were still the result of the actions of a relatively active government type, whereas in district VIII a shift from an active to a passive government type had taken place. Similar variations across administrative districts existed in Amsterdam too: Buurt Negen still exemplifies the active government type, whereas de Pijp experienced a gradual shift from an active to a more passive government type.

The challenge for future undertakings will be to find a way to disentangle between effects induced by more active and passive government types from the data produced. As already noted, in the two city-contexts under study the situation is different in this respect: whereas in Amsterdam the highly detailed dataset made such divisions difficult, the less detailed data of Budapest enabled more general analyses, and thereby also a rudimentary division between the two effects. Furthermore, when the aim is to measure the price and quality effects caused by either type of institutional ideal situation, that is to say, the active or passive government type, it may be that the level of detail has to be limited to only detecting increases over broadly defined price categories. Besides, the time-period of the study undoubtedly ought to be longer, up to twenty years, and even then it may not be possible to find out whether price levels are different in two areas one of which is revitalised and the other which is not.[6]

Discussion on housing market process in urban renewal areas

This chapter has documented an attempt to ascertain the key tendencies in disentangling *where* and *why* certain changes in the built environment happen. This conceptualisation involves three interlinked elements: one, the behaviour of the producers (and supply in the broad sense) in relation to that of the consumers (demand); two, the more active role of government in relation to that of a more passive government role; and three, the motivations of profit in relation to those of sustainability.

What kind of generalisations can be made from the main findings of the case studies? First of all, in Budapest the situation was that differences between the two areas were 'black and white'; this becomes clear, even though inferior data were used. In Amsterdam the situation was such that differences between the two areas were subtle and part of a total picture; the dimensions under study cannot be isolated easily even from high quality data. Thus the paradox here is that the more and better the data, the less clear the result becomes.

Moreover, it can be concluded that, when using various explanatory variables for a certain price increase and/or for a certain quality increase, tenure change from renter to ownership mattered more in the Budapest case, as the Amsterdam housing market to a great extent was dominated by social rental apartments – this applies for these two specific urban regeneration areas as well as for the city as a whole. On the other hand, this quick tenure change process was largely supply driven in Budapest (cf. Kovács, 2009). Perhaps in parts of Amsterdam such as in de Pijp it is realistic that on the demand side the driving force was the possibility of the prospective owners to get tax deductions by changing tenure from rental to ownership – given a situation where the local government had become a more passive player. There is, in fact, evidence on the country level that buyers got tax incentives to buy apartments and housing associations got incentives to sell their stock (Aalbers, 2008, p. 157). This would imply both demand and supply led mechanisms of institutional change driven by incentives and eventually a confirmation of the value gap theory. On the supply side, in turn, two kinds of drivers were at play in Amsterdam: on the one hand, the pure profit motivations of the developers and builders, and possibly also the actions of the local government; on the other hand, how the planning system could guarantee a long-term economic and other sustainability of the area. In a sense, the evidence compiled fits stereotypical 'East vs. West' differences in economic and cultural outcomes.

The findings indicate that, in the Amsterdam context of urban regeneration in general and the context of housing development in particular, the development strategies are more sustainable environmentally, socially and economically; more demand driven. And, social housing still has a dominant position, still influenced by a more active government than in Budapest, although we can identify a government agenda in the latter case too. There is furthermore a causality between these elements insofar as real estate sustainability is dependent on *both* increased demand influence – when products need to be targeted for the right consumer groups, bar urban sprawl, environmental hazards, price inflation or other harmful effects – and having strong housing policies (to assure that the society's interest is considered too). Policies can also be designed as to stimulate demand (e.g. mortgage interest tax relief), as illustrated schematically in Figure 7.8. Overall the conclusion must be that what really matters is the active local government – good local governance in whatever form (e.g. slum clearance, PPP) is an important precondition in this context.

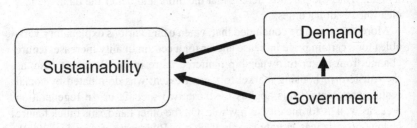

Figure 7.8 Dependence of sustainability on demand and government.

Notes

1 This chapter builds on three conference presentations by the same author (EUGEO Amsterdam 2007; RSA Prague 2008; EUGEO Bratislava 2009).
2 The Marxist tradition of geographers constitutes an earlier attempt of such analysis. To make it explicit: in the current study the economic theory framework is *not* related to Marxism; it is rather heterodox and inspired by institutionalists such as Thorstein Veblen and behaviourists such as Herbert A. Simon.
3 The empirical material presented here is the same as in Kauko (2009a). In the present contribution a sustainability view of economic geography is elaborated, which was not the case in the prior article.
4 We must also note that in the middle part of district VIII for a long time nothing happened; but prices went up; that was still the situation when I began this work in 2004 and got time-series data of house prices (1998–2002). However, since then enormous changes have occurred especially in the *Corvin* neighbourhood – block-wise developments; homogenisation. Thus the data presented here are outdated but serves well for illustrating general trends and tendencies.
5 This is as such a socialist type model of city development!
6 I am indebted to Professor Alan Murie for this point, which he brought up during the session of the EUGEO conference in Amsterdam in August 2007.

8   Concluding discussion

Identifiable sustainability elements in a real estate context

Economic sustainability and real estate

As a result of a near universal trend in social values towards an economical use of resources, avoidance of pollution and related aspirations, in many urban areas real estate investments are now considered in a more sustainable framework. The bottom line is that a sustainable place generates a competitive advantage over unsustainable places. Here real estate and housing is an important element in defining what a sustainable place is, because jobs and homes require their physical settings. The question is about the physical, social and economic sustainability of investments. The possibility, feasibility and necessity for either new development or refurbishment of the building stock depends on the character of the area within the city, the city itself, and the institutional setting where investment takes place. Furthermore it is about the quality-of-life (QOL) of the people occupying the dwellings and their daily living environment. While this topic has business and social policy relevance, academic work on defining the relevant economic, physical and social assessment criteria is yet speculative, when standard definitions as well as relevant data are lacking.

In this book about real estate development in relation to socio-economic sustainability metrics and criteria two basic questions form the core of the research problem:

1   What is socio-economically sustainable urban real estate – thus, how can it be defined?
2   How to bring a change towards socio-economically sustainable urban real estate – in other words, how to legitimate and execute the process?

The solution proposed here lies in the economic (including financial and socio-economic) sustainability concept. Insofar as sustainable real estate development is concerned, urban residential and office development projects may be summarised into two categories: (1) if only *profit* is considered, the project is unsustainable in the long run; (2) if *sustainability* related motives dominate normal profits are reaped and the remaining margins are fed back onto the use value of the project. As argued at the outset, social, cultural and environmental sustainability are here treated as secondary goals that can be financed only after the economic side is solid. Thus, these issues are subordinate to the economic growth necessity and only to be implemented if the economic management is apt to provide means to improve the other sustainability considerations in a given project or plan. This then is a private investment and market data driven process but with the local government (or other public body) implementing 'smart regulations' to safeguard that the secondary sustainability goals are met. This does not, however, conform to mainstream thinking about sustainable development and herein lies the originality of this study.

In other words, a novel framework is proposed on the basis of the concept of economic sustainability that is argued to connect with the financial crisis, credit crunch and housing bubble discourse – it is to note that, traditionally, housing market modelling (quantitative, economic measure) and sustainable development paradigms (qualitative, multidimensional including non-economic measure) are treated as incompatible. One of the reasons is that defining a sustainable housing market depends on long-term processes and often partial criteria. In the previous chapters this idea has been discussed and concretised using different lines of argumentation. For residential property at least three issues are pertinent: (1) the price in relation to quality, (2) the price in relation to income, and (3) the diversity of the product insofar as this would trigger innovation – that is to say, 'not putting all the eggs in the same basket'. The empirical parts of the study concern the development of house prices in relation to incomes in Trondheim, Norway (in the period 1993–2008); the classification of the development of house prices in relation to subjective quality in various parts of Amsterdam, the Netherlands (during 1986–2002); and the price development in different parts of Budapest and Amsterdam that are undergoing urban renewal (mainly during the late 1990s). Despite the limitations of the data, in terms of its time and coverage, the findings have a variety of directly significant implications for sustainability assessment in a real estate context.

The analysis pertains to residential property but can be generalised to other kinds of urban property types as well. First, the study proposes some general definitions and criteria for creating a sustainable built environment. Second, among several vaguely defined, non-standard and competing definitions for economic sustainability, the study applies a rather pragmatic *value stability* concept as a basis for a narrowly defined, but manageable methodology. The study yields a number of interesting outcomes regarding prices in relation to measures of quality, affordability and diversity:

- As grossly substandard level of housing is unacceptable for health and safety reasons, the quality (largely a subjective indicator though) ought to develop in the same direction and with the same pace as the price level; this is about the site and building specific attributes as well as the characteristics of the surrounding environment, neighbourhood and the city as a whole. What is socially sustainable is not necessary economically sustainable, and *vice versa*.
- As it is not sufficient with high quality unless people cannot afford to buy (or rent) it, the affordability (often approximated as net income) of the dwelling also ought to develop in the same direction and with the same pace as the price level. It may be found that some of the wealthiest areas are also among the economically least sustainable ones.
- Even if the quality and affordability criteria are fulfilled, it is not sufficient for value stability (and hence economic sustainability) unless there is a *wide enough range* (i.e. product variety generated for most apt selections to be made) of different quality and affordability levels on the market. This is because the drivers of sustainability: production technology, community governance as well as consumption fashions, all tend to change quickly and then it is vital not to have neglected any specific property/housing package even if it may seem marginal at some stage. (Or put it differently, if a potential market trend setter or other innovation in terms of quality or affordability is not recognised this will have harmful impacts for the evolution of the property portfolio in terms of its value stability.) When examining the price development for different dwelling types and locations that are assumed affected and unaffected by urban renewal projects it may be that some of the examined market segments show a more heterogeneous development in terms of mix than in other areas. Given the previous postulations, this in turn gives a basis for the argument that the more heterogeneous area, in this respect, also is more economically sustainable than the more homogeneous case.

Finding a valid sustainability metrics

In the quantitative parts (documented in Chapters 5–6), the first task was to analyse the development of house prices in relation to income in Trondheim 1993–2007.[1] The second aim was to classify the citywide residential property market to *sustainable or unsustainable* based on price in relation to the quality of the dwelling and environment; that study used a dataset from Amsterdam 1986–2002, with sales prices and a number of attributes including subjective environmental and housing quality as input variables. The potential applicability of these studies can be seen in their pragmatic approach to the construction of property price based sustainability indicators. Adding considerations of valuation automata to the propositions by Sayce and colleagues (2007) discussed in Chapter 3 results in a procedure where valuation automata is applied consistently on local house prices and other complex databases over several years, in order to develop various economic sustainability metrics, which in turn is vital for setting the appropriate incentives for sustainable real estate investments and government policies. This principle is illustrated diagrammatically in Figure 8.1. In the AVM other kinds of quantifiable data such as income or real estate quality indicators may be worth combining with the price data. Subsequently, less tangible criteria and indicators can be incorporated to the outcome of the automated modelling of real estate value. Finally, the resulting metrics based on long-term and partial criteria makes a convenient basis for real world decisions concerning economic sustainability.

Due to its pragmatic and non-linear nature the quasi-dynamic approach created in Chapters 4–6 seems promising. Given the increasing significance

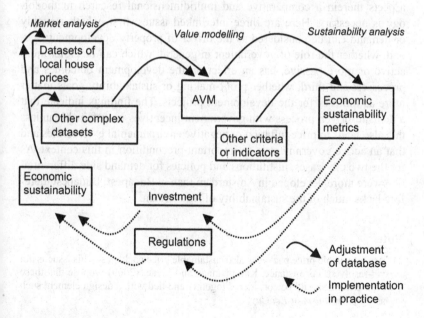

Figure 8.1 The proposed sequence of procedures for the development and implementation of an economic sustainability metrics.

of the sustainability discourse within planning, the findings also have policy implications: using innovative empirical housing market modelling as an analytic tool enables planning for sustainable land use and governance on an urban and regional scale. A further, more practical implication of all this is that all properties could in principle be 'sustainable' and this is not only with regard to the physical aspects but with regard to socio-economic aspects too. That said, for some academics this approach might seem too different from an economic mainstream approach (see Lorenz et al., 2008). Therefore one is well advised to tread with caution. As stated at the outset the ambition level probably should be reduced from developing tools to capturing the turning points in market development and eliminating the risk involved to tools that merely manages the unfavourable situation. Data quality poses further requirements to this project. While this obviously is not 'rocket science', an intellectual breakthrough could be achieved by reaching out towards discussions evolving elsewhere such as those pertaining to the definitions and criteria for creating a sustainable built environment.

In the last empirical module (documented in Chapter 7) it was shown how, when examining urban renewal processes, the choice of the location is crucial. Furthermore, in order to ascertain institutional and behavioural aspects therein a comparative and multidimensional research methodology is necessary. Here are three interlinked issues: first, whether supply or demand can be considered as the driver of property development; second, whether the role of government in general, which can be either more active or more passive, has an effect on the development outcomes and processes; and third, whether profit-making or sustainability goals matter more as motives for the development projects. The findings indicate that a demand-driven process with government incentives is sustainable unless this also leads to price inflation or negative environmental externalities and that an active government is an important precondition in this context. As for the two city-cases, institutions and policies for demand side differentiation were more developed in Amsterdam than in Budapest. Budapest therefore lacks much of the sustainability of Amsterdam.

Note

1   Are affordable price packages also sustainable price packages? This issue is not yet resolved. For instance, Karuppannan and Sivam (2009) propose that these both criteria should be considered together and tied with a design element such as *New Urbanism* or *EcoCity*.

Epilogue[1]

At the moment our civilisation is facing a variety of challenges related to sustainable development. As also this book has shown, in these difficult times we are stretched to go far beyond our comfort zones. Problems in relation to inequality and use of resources can be solved only when we are open to new possibilities. It is therefore our duty to be aware of what lies 'out there' – different traditional and state-of-the-art research settings and theoretical-methodological traditions need to be examined every now and then. In this examination, even the sustainable development paradigm is not immune from critique. We must not forget that it is community values that are at stake when the times are strongly encouraging narrow minded individualistic thinking. Our ultimate goals must therefore be defined in terms of how we intend to manage keeping the urban environment habitable for future generations too.

Neither theoretical nor empirical studies work effectively towards these goals on their own. Theory is based on idealised settings, thus extremes, and often not useful in the daily life; empiry in turn is usually based on average figures and therefore not interesting for academic audiences, even if useful in daily life. 'In theory there is no difference between theory and practice, but in practice there is' as the proverb goes. It is, however, by combining the two sides that we can move the research frontier forward. This is particularly relevant in hitherto underexplored topics such as the socio-economic side of urban real estate sustainability, and this is regardless of exact definitions made.

Note

1 Inspiration to these afterthoughts arose after attending the *8th workshop on planning and evaluation 'Evaluation in Integrated Land-use management – Towards area-oriented and place-based evaluation for infrastructure and spatial projects'*, 13–15 March 2013, Groningen.

Appendix 1

The SOM

In a generic sense, the SOM represents a neural network technique suitable for data mining as well as clustering and classification. Having both numerical and visual properties, the clustering and classification technique applied in this study, the SOM, fits well here too (see also Kauko, 2002, 2004c, 2005, 2006a, 2009b, c). According to the basic idea of the SOM algorithms an n-dimensional input data matrix, where n is the number of variables, is compressed to an output, where the array of nodes comprises two dimensions and numerical values of each node (feature map). Each *layer* in the feature map represents one variable. For every node, 'typical values' are computed for each layer. Also the empty nodes (i.e. the nodes which do not 'win' any observations and remain without label) obtain 'typical values' (Kohonen, 1995).

The map is generated in three main steps: first, initialisation; then, selection of network parameters; and finally, calibration of a stabilised map. The following technical terminology is necessary in order to understand the procedures carried out:

- *Nodes* defined through the x and y map coordinates, indicate the maximum number of potential clusters. (Note that these are not geographical coordinates but coordinates that determine the location of the node in the SOM output matrix.) To illustrate with the extreme example: if we define only one node, we obtain one cluster that corresponds to all N observations; if the number of observations is the same as the number of nodes, there are N possibilities: we might obtain N clusters with one observation each, only one cluster with N observations surrounded by N−1 empty nodes, or all possibilities in between these two extremes.
- *Dimensions* may refer to three different concepts: (1) Size (i.e. number of nodes) and shape[1] of the map (the coordinates x and y of the nodes). Here we define map dimensions as small as possible in order

to retain tractability, as defining a larger map makes the analysis fuzzier. This parameter is decidedly *ad hoc*: for example, here we defined 6 and 12 node variants; Kauko (2005), however, defined 192 nodes. (2) The 11 input variables shown in Table 6.1 are referred to as original dimensions. (3) The expected factors, the interpretation of which is driven by theory, represent principle components extracted from the original dimensions, and are also linked to the map dimensions.

* *Clusters* on the map can be defined in two ways: They may be identified based on the resulting fuzzy patterns on the map. In the prior study Kauko (2005) found five such clusters from a map of 192 nodes. Alternatively, the clustering can also be defined directly based on the nodes in the map, which then requires a smaller map. However, if the map is too small (i.e. fewer than six nodes), the resolution is insufficient to discern all five expected segments. Therefore, here a map of six nodes is selected as a starting point (see Figure 4.1).

Depending on how stable, well-structured and valid a map we require, we also need to define the learning process in terms of the following parameters that direct the process of map adjustment for each comparison between input observation and output response (iteration):[2]

* The number of iterations to run the algorithm we consider necessary,
* The sensitivity and extent of the 'active' part of the stabilising map to the adjustment prompted by each new iteration.

We may test the robustness of the outcome by defining a different map. The map size affects the resolution: a 4 by 3 map reveals more classes than a 3 by 2 map; at the same time, however, the larger map shows more noise between the categories. We can note that the technique itself is easy to use, as it, like other neural network techniques, 'eats' all kinds of data, and even allows for missing entries in the input matrix. On the other hand, it is really when the output matrix is obtained that the analysis begins in earnest: to try to relate the result to relevant external knowledge – empirical context or more general theoretical considerations. This is then considered a partly qualitative approach. To add some more rigour we may therefore apply various numerical procedures.

While the Kohonen Map is a more pragmatic and holistic method than the more traditional methodologies, it has been criticised for being too much of a black box approach (see Kauko, 2002). For relevant prior geodemographic classification work, Openshaw et al. (1994) use the SOM on British census data, and Hatzichristos (2004) carried out a demographic classification of Athens, Greece, using the SOM in combination with fuzzy logic. Prior studies on clustering housing market areas/regions using the SOM have been carried out by the author (see e.g. Kauko, 2005).

The essence of the SOM is about estimating the distribution of the dataset, when the output of the SOM can be associated with the geographical maps. This means two kinds of dispersions: one is about nearby situated locations that have different attributes; the other is about far away situated locations that have similar attributes. Unfortunately, few studies to date demonstrate the novelty of this aspect within a socio-economic urban context. In one such contribution, however, Spielman and Thiel (2007) explain how the data mining property of SOM can be used for social area analysis, more specifically, to demonstrate how similar places are found in different parts of one and the same city. The idea is to identify how geographic location is related to social similarity and difference in terms of the dimensions defined by a matrix of seventy-nine attributes measured at the course level of the 2217 census tracts in New York City (cf. Webber, 2007; Kauko, 2009b). Spielman and Thiel conclude by discussing the benefits and drawbacks of the method: while it is positive that many kinds of application areas are possible for the SOM, the problem is that the SOM shares many of the limitations of factor analysis and geo-demographic clustering techniques including the absence of a direct theoretical guidance for selection of variables.

Specifically, the SOM is also used as a method for pattern seeking and coarse classification of statistical housing market districts. While standard procedures use too stringent deterministic rules that may hide informative patterns, in this work we apply the Kohonen Map (self-organising map, SOM). The Kohonen Map is a neural network technique that can be interpreted as a close relative to the established k-means clustering technique.[3]

Valid and reliable data together with sound statistical techniques do matter in the process of monitoring and analysing residential patterns and housing market diversity. Normally, a set of variables is selected and given meanings using data-reduction (principal component analysis and factor analysis) and/or clustering methods. When classifying and assessing housing areas, the minimum requirement for such a system is a solid typology together with a spatial component; even better, if it allows utilising fuzzy and non-linear modelling concepts. Last, it is fair to say that, with the SOM one is able to show several handy applications, but it cannot be argued that it is the best of this type of technique.

Notes

1 When aiming at a stable map, it is better to choose the x and y dimensions of different size (such as 3×2, 4×3 and 16×12) than to define a squared shape (such as 3×3, 4×4 or 15×15).

2 This is a simulation approach, where the map learns by trial-and-error until the organisation has reached a sufficient convergence. The network parameters are compulsory and more or less *ad hoc* choices to make by the analyst in this application, which is based on the SOM_PAK software (see Kohonen et al, 1996). The issue is also about feasibility as selecting a longer running time and 'more optimal' parameters result in a slower procedure.

3 Compared to the *k*-means technique, the SOM enables to freely select the most appropriate clustering, as it displays the 'real' distances between the identified clusters in a fuzzy and non-linear fashion. To give an example from an earlier study by one of the authors (Kauko, 2002, on Helsinki), the SOM and the *k*-means discerned the same four clusters, but when the number of potential clusters was reduced to three, the methods identified a different clustering. When the outcome was validated based on external knowledge, it turned out that the SOM had captured the true segmentation, whereas the *k*-means clustering had merged totally different areas into one and the same segment and split a homogeneous area instead.

Appendix 2

Feature maps (SOM output) of Trondheim, 1993–2007

Each map layer below shows the variation in price per annual average income for the whole city area. The label denotes ward and is for identification only.

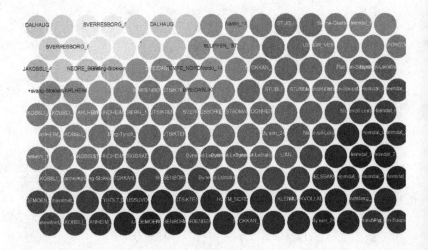

1993

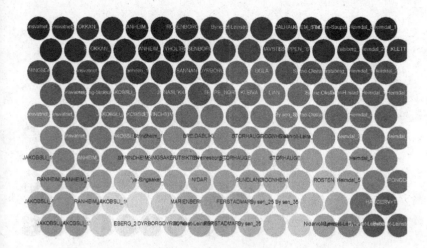

1994

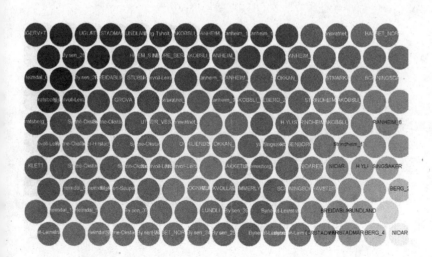

1995

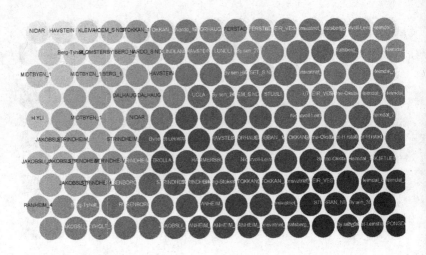

1996

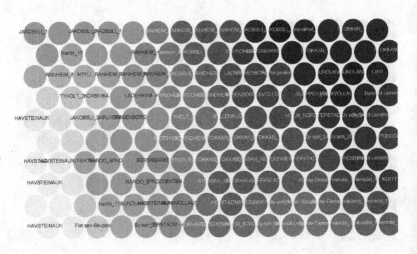

1997

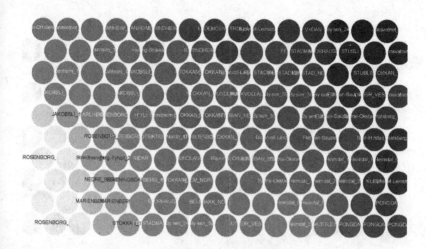

1998

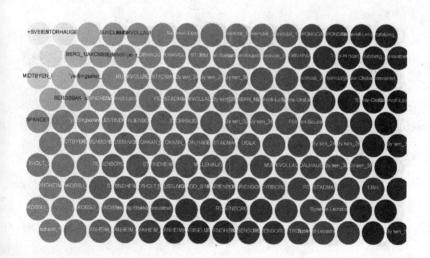

1999

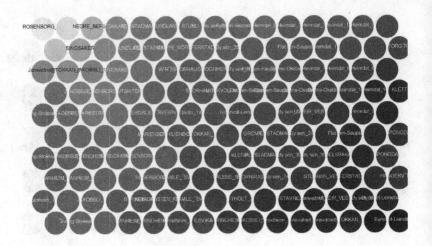

2000

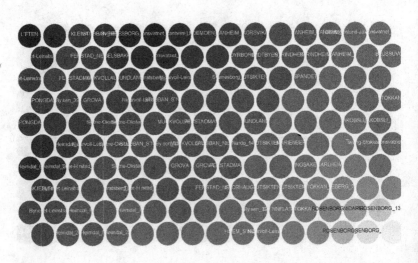

2001

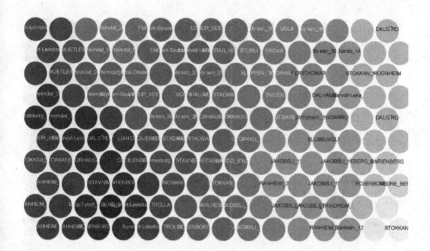

2002

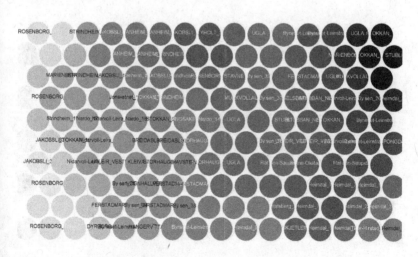

2003

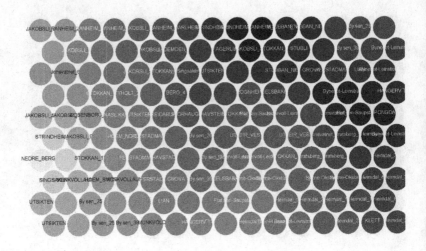

2004

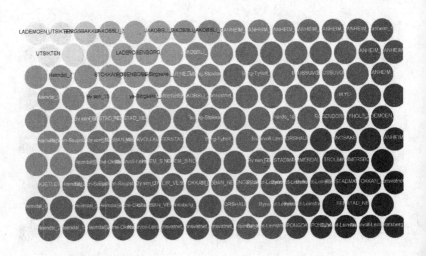

2005

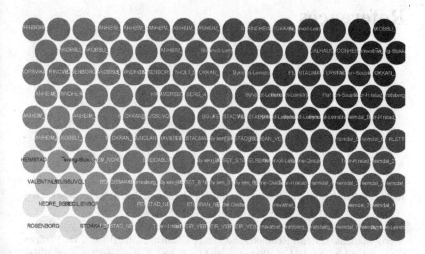

2006

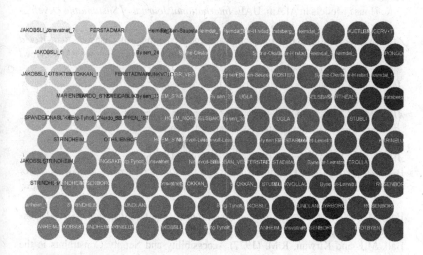

2007

References

Aalbers, M.B. (2008) The Financialization of Home and the Mortgage Market Crisis. *Competition & Change*, 12(2), June, pp. 148–166.

Adams, D., Watkins, C. and White, M. (2005) Planning, Public Policy and Property Markets: Current Relations and Future Challenges. In: Adams, D., Watkins, C. and White, M. (Eds.): *Planning, Public Policy & Property Markets*. Wiley-Blackwell, Oxford, pp. 239–251.

Addae-Dapaah, K., Hiang, L.K. and Shi, N.Y. (2009) Sustainability of Sustainable Real Property Development. *Journal of Sustainable Real Estate*, 1(1), pp. 203–225.

Ahlmann, H.W. (1917) De norska städernas geografiska förutsättningar (in Swedish). *YMER*, 3, pp. 17–299.

Ahmed, K.G. (2011) Evaluation of Social and Cultural Sustainabilty in Typical Public House Models in Al Ain, UAE. *International Journal of Sustainable Development Planning*, 6(1), pp. 49–80.

Andrews, M. (2008) Risk, Inequality and the Economics of Disaster. *Real-World Economics Review*, 45, 15 March, pp. 2–9. http://www.paecon.net/PAEReview/issue45/Andrews45.pdf.

AVM News (2008a) e-newsletter, issue July, August.

AVM News (2008b) e-newsletter, issue September, October.

AVM News (2008c) e-newsletter, issue November, December.

AVM News (2009) e-newsletter, issue January, February.

AVM News (2011) e-newsletter, issue March, April.

AVM News (2014) e-newsletter, issue September, October.

Baker, D. (2008) The Housing Bubble and the Financial Crisis. *Real-World Economics Review*, 46, 20 May 2008, pp. 73–81. http://www.paecon.net/PAEReview/issue46/Baker46.pdf.

Ball, M. (2006) *Markets & Institutions in Real Estate Construction*. Wiley-Blackwell, Oxford.

Ball, M.J. and Kirwan, R.M. (1977) Accessibility and Supply Constraints in the Urban Housing Market. *Urban Studies*, 14, pp. 11–32.

Bénabou, R. and Tirole, J. (2009) *Individual and Corporate Social Responsibility*. Unpublished manuscript.

Bitušiková, A. and Luther, D. (2010) Sustainable Diversity and Public Space in the City of Bratislava, Slovakia. *Anthropological Notebooks*, 16(2), pp. 5–18.

Bogliotti, C. and Spangenberg, J.H. (2006) A Conceptual Model to Frame Goals of Sustainable Development. *International Journal of Sustainable Development Planning*, 1(4), pp. 381–398.

Bontje, M. (2004) From Suburbia to Post-Suburbia in the Netherlands: Potentials and Threats for Sustainable Regional Development. *Journal of Housing and the Built Environment*, 19, pp. 25–47.

Bramley, G., Dempsey, N., Power, S., Brown, C. and Watkins, D. (2009) Social Sustainability and Urban Form: Evidence from Five British Cities. *Environment and Planning A*, 41, pp. 2125–2142.

Bramley, G., Leishman, C. and Watkins, D. (2008) Understanding Neighbourhood Housing Markets: Regional Context, Disequilibrium, Sub-Markets and Supply. *Housing Studies*, 23(2), pp. 179–212.

Bramley, G. and Power, S. (2009) Urban form and Social Sustainability: The Role of Density and Housing Type. *Environment and Planning B: Planning and Design*, 36(9), pp. 30–48.

Brattbakk, I. and Hansen, T. (2004) Post-War Large Housing Estates in Norway – Well-Kept Residential Areas Still Stigmatised? *Journal of Housing and the Built Environment*, 19, pp. 311–3323.

Brattbakk, I., Jørgensen, S. and Dale, B. (2000) *Stabilitet eller Endring? Levekårsutvikling i Trondheims boområder på 1990-tallet* (in Norwegian). NTNU, Trondheim.

Bromley, D.W. (2007) Environmental Regulations and the Problem of Sustainability: Moving beyond "Market Failure". *Ecological Economics*, 63, pp. 676–683.

Bryson, J.R. (1997) Obsolescence and the Process of Creative Reconstruction. *Urban Studies*, 34(8), pp. 1439–1458.

Bryson, J.R. and Lombardi, R. (2009) Balancing Product and Process Sustainability against Business Profitability: Sustainability as a Competitive Strategy in the Property Development Process. *Business Strategy and the Environment*, 18(2), pp. 97–107.

Budd, L. and Hirmis, A. (2004) Conceptual Framework for regional Competitiveness. *Regional Studies*, 38(9), pp. 1015–1028.

Cajias, M. and Bienert, S. (2011) Does Sustainability Pay Off for European Listed Real Estate Companies? The Dynamics between Risk and Provision of Responsible Information. *Journal of Sustainable Real Estate*, 3(1), pp. 211–231.

Cameron, S. (2006) From Low Demand to Rising Aspirations: Housing Market Renewal within Regional and Neighbourhood Regeneration Policy. *Housing Studies*, 21(1), pp. 3–16.

Carlson, E. (1998) Real Estate Investment Appraisal of Land Properties Using SOM. In: Deboeck, G. and Kohonen, T. (Eds.): *Visual Explorations in Finance with Self-Organizing Maps*. Springer, New York, pp. 117–127.

Cheshire, P. (2005) Unpriced Regulatory Risk and the Competition of Rules: Unconsidered Implications of Land Use Planning. *Journal of Property Research*, 22(2–3), June–September, pp. 225–244.

Clark, E. (1987) The Rent Gap and Urban Change. Case Studies in Malmö 1860–1985. Meddelanden från Lunds Universitets geografiska Institutionen, avhandlingar 101, Lund University Press, Lund.

Colantonio, A. and Dixon, T. (2011) *Urban Regeneration and Social Sustainability: Best Practice from European Cities*. RICS Research, Wiley-Blackwell, Oxford.

Cook, I.R. and Swyngedouw, E. (2014) Cities, Nature and Sustainability. In: Paddison, R. and McCann, E. (Eds.): *Cities & Social Change*. Sage, London, pp. 168–185.

Couch, C. and Fraser, C. (2003) Introduction: The European Context and Theoretical Framework. In: Couch, C., Fraser, C. and Percy, S. (Eds.): *Urban Regeneration in Europe*, Real Estate Issues. Wiley-Blackwell, Oxford, pp. 1–16.

Cox, J., Fell, D. and Thurstain-Goodwin, M. (2002) *Red Man, Green Man, Performance Indicators for Urban Sustainability*. RICS Foundation London.

Coyle, S.J. (2011) The Process of Transformation. Sustainable Plan-Making. In: Coyle, S.J. and Duany, A. (Eds.): *Sustainable and Resilient Communities: A Comprehensive Action Plan for Towns, Cities, and Regions*. John Wiley and Sons, Hoboken, NJ, April, pp. 25–31.

d'Amato, M. and Kauko, T. (2008) Property Market Classification and Mass Appraisal Methodology. In: Kauko, T. and d'Amato, M. (Eds.): *Mass Appraisal Methods – An International Perspective for Property Valuers*. RICS Series, Wiley-Blackwell, Oxford, pp. 280–303.

D'Arcy, E. and Keogh, G. (1998) Territorial Competition and Property Market Process: An Exploratory Analysis. *Urban Studies*, 35, pp. 1215–1230.

Davoudi, S., Hall, P. and Power, A. (2008) Key Issues for Planning Futures and the Way Forward. *21st Century Society*, 3(3), pp. 229–247.

Deng, Y., Li, Z. and Quigley, J.M. (2010) Economic Returns to Energy-Efficient Investments in the Housing Market: Evidence from Singapore. IRES Working Paper Series IRES2010–008, August.

Dent, P. and Dalton, G. (2010) Climate Change and Professional Surveying Programmes of Study. *International Journal of Sustainability in Higher Education*, 11(3), pp. 274–291.

Dixon, T. (2007) The Property Development Industry and Sustainable Urban Brownfield Regeneration in England: An Analysis of Case Studies in Thames Gateway and Greater Manchester. *Urban Studies*, 44(12), pp. 2379–2400.

Dixon, T., Thompson, B., McAllister, P. and Snow, J. (2005) *Real Estate & the New Economy. The Impact of Information and Communications Technology*. Wiley-Blackwell, Oxford.

Downie, M.L. and Robson, G. (2007) *Automated Valuation Models: An International Perspective*. The Council of Mortgage Lenders (CML), London, October.

du Plessis, C. and Cole, R.J. (2011) Motivating Change: Shifting the Paradigm. *Building Research & Information*, 39(5), pp. 436–449.

EC (2010) World and European Sustainable Cities. Insights from EU Research. European Commission, European Research Area, Socio-Economic Sciences and Humanities. Brussels.

Égert, B. and Mihaljek, D. (2008) Determinants of House Prices in Central and Eastern Europe. Czech National Bank Working Paper. http://www.eukn.org/eukn/themes/Urban_Policy/Housing/Housing_management/Housing_finance/House_prices/House-Prices-in-Central-and-Eastern-Europe_1011.html [accessed 25 May, 2009].

Eichholz, P., Kok, N. and Quigley, J. (2009) *Why Do Companies Rent Green? Real Property and Corporate Social Responsibility.* RICS Research, London.

Eichholz, P., Kok, N. and Quigley, J. (2010) Doing Well by Doing Good? Green Office Buildings. *American Economic Review,* 100, December, pp. 2492–2509.

Ellison, L., Sayce, S. and Smith, J. (2007) Socially Responsible Property Investment: Quantifying the Relationship between Sustainability and Investment Property Worth. *Journal of Property Research,* 24(3), pp. 191–219.

ESPON (2010) First ESPON 2013 Synthesis Report. New Evidence on Smart, Sustainable and Inclusive Territories. http://www.espon.eu [accessed 29 June, 2011].

Evans, A.W. and Hartwich, O.M. (2005) *Bigger Better Faster More. Why Some Countries Plan Better than Others.* Policy Exchange, London. http://www.policyexchange.org.uk/images/publications/pdfs/pub_39_-_full_publication.pdf [accessed 25 May, 2009].

Fahy, F. and Ó Cinnéide, M. (2008) The Reality of the Locality: Exploring Spatial Aspects of Quality of Life in Galway City, Ireland. *International Journal of Sustainable Development Planning,* 3(1), pp. 29–44.

Farkas, J., Giczi, J. and Székely, G. (2004) *Housing Conditions 1999–2003.* Hungarian Central Statistical Office. http://mek.oszk.hu/06900/06979/06979.pdf [accessed 22 September 2011].

Fisher, R. (2010) The Role of Local Government in Designing Sustainable Cities. Presentation at the EU Sustainable Energy Week (EUSEW), 22–26 March, Brussels.

Földi, Z. (2006) *Neighbourhood Dynamics in Inner-Budapest. A Realist Approach.* Nederlands Geographical Studies 350. Utrecht University, Utrecht, 2006.

Fossett, M. and Waren, W. (2005) Overlooked Implications of Ethnic Preferences for Residential Segregation in Agent-Based Models. *Urban Studies,* 42(11), pp. 1893–1917.

Foxon, T.J., Köhler, J., Michie, J. and Oughton, C. (2012) Towards a New Complexity Economics for Sustainability. *Cambridge Journal of Economics,* Advance access, 23 November, pp. 1–22.

Francke, M. (2010) Casametrics – The Art of Modelling and Forecasting the Market Value of Houses. English summary of inaugural lecture, University of Amsterdam, 4 February.

Fuerst, F. and McAllister, P. (2011) Green Noise or Green Value? Measuring the Effects of Environmental Certification on Office Values. *Real Estate Economics,* 39(1), pp. 45–69.

Galuppo, L.A. and Tu, C. (2010) Capital Markets and Sustainable Real Estate: What Are the Perceived Risks and Barriers? *Journal of Sustainable Real Estate,* 2(1), pp. 143–159.

Ganser, R. and Williams, K. (2007) Are We Using the Right Targets? Evidence from England and Germany. *European Planning Studies,* 15(5), pp. 603–622.

Geltner, D. and de Neufville, R. (2012a) Uncertainty, Flexibility, Valuation and Design: How 21st Century Information and Knowledge Can Improve 21st Century Urban Development (Part 1 of 2). *Pacific Rim Property Research Journal,* 18(3), September, pp. 231–249.

Geltner, D. and de Neufville, R. (2012b) Uncertainty, Flexibility, Valuation and Design: How 21st Century Information and Knowledge Can Improve 21st

Century Urban Development (Part 2 of 2). *Pacific Rim Property Research Journal*, 18(3), September, pp. 251–276.

Glaeser, E. (2011) *The Triumph of the City*. Macmillan/Pan-Books, London.

Glaeser, E.L. and Resseger, M.G. (2009) The Complementarity between Cities and Skills. NBER Working Paper No. 15103, June.

Goering, J. (2009) Sustainable Real Estate Development: The Dynamics of Market Penetration. *Journal of Sustainable Real Estate*, 1(1), pp. 167–201.

González, M.A.S. and Kern, A.P. (2007) A Framework to Sustainable Renewal of Existing Building Stock in Brasil. SB07 conference, Lisbon.

Greenfield, A. (2013) *Against the Smart City*. Do projects, New York City.

Grigsby, W., Baraty, M., Galster, G. and Maclennan, D. (1987) The Dynamics of Neighbourhood Change and Decline. *Progress in Planning*, 28(1), pp. 1–76.

Hamnett, C. (2003) *Unequal City. London in the Global Arena*. Routledge, London.

Harrison, D.M. and Seiler, M.J. (2011) The Political Economy of Green Industrial Warehouses. *Journal of Sustainable Real Estate*, 3(1), pp. 44–67.

Hatzichristos, T. (2004) Delineation of Demographic Regions with GIS and Computational Intelligence. *Environment and Planning B*, 31, pp. 39–49.

Hegedűs, G. (2011) Geographical Analysis of Gated Communities and Their Challenges for Urban Development in Hungary. "Summary and suggestions" and "Theses of PhD dissertation", University of Szeged, Department of Economic and Social Geography, May, Szeged.

Hemphill, L., Berry, J. and McGreal, S. (2004a) An Indicator-Based Approach to Measuring Sustainable Urban Regeneration Performance: Part 1, Conceptual Foundations and Methodological Framework. *Urban Studies*, 41, pp. 725–755.

Hemphill, L., Berry, J. and McGreal, S. (2004b) An Indicator-Based Approach to Measuring Sustainable Urban Regeneration Performance: Part 2, Empirical Evaluation and Case-Study Analysis. *Urban Studies*, 41, pp. 757–772.

Hill, S., Lorenz, D., Dent, P. and Lützkendorf, T. (2011) Built Environment Professionalism in a Changing Economy. Paper for the SBO conference.

Hobson, K. (2008) Reasons to Be Cheerful: Thinking Sustainably in a (Climate) Changing World. *Geography Compass*, 2/1, pp. 199–214.

Holt-Jensen, A. (2009) The Market Oriented and Privatized Housing Culture in Norway. In: Holt-Jensen, A. and Pollock, E. (Eds.): *Urban Sustainability and Governance*. Nova Science Publishers, New York (NY). pp. 119–131.

Holzmarkt (2014) *Concept & Architecture* [in German]. http://www.holzmarkt.com [accessed 17 September, 2014].

Hoornweg, D., Sugar, L. and Trejos Gomez, C. (2011) Cities and Greenhouse Gas Emissions: Moving Forward. *Environment & Urbanization*, XX(X), pp. 1–21.

Hu, H. (2014) Sustainable Living in a Chinese City. Analysis and Support for Market-Conscious Urban Planning. PhD Thesis, Utrecht University.

Jackson, J. (2009) How Risky Are Sustainable Real Estate Projects? An Evaluation of LEED and ENERGY STAT Development Options. *Journal of Sustainable Real Estate*, 1(1), pp. 91–106.

Jäger, J. (2003) Urban Land Rent Theory: A Regulationist Perspective. *International Journal of Urban and Regional Research*, 27(2), pp. 233–249.

Jones, C., Leishman, C. and MacDonald, C. (2009) Sustainable Urban Form and Residential Development Viability. *Environment and Planning A*, 28 April, pp. 1–24.

Jones, C. and Watkins, C. (1996) Urban Regeneration and Sustainable Markets. *Urban Studies*, 33(7), pp. 1129–1140.

Joss, S. (2011) Eco-Cities: The Mainstreaming of Urban Sustainability – Key Characteristics and Driving Factors. *International Journal of Sustainable Development Planning*, 6(3), p. 268.

JOSRE is published by ARES in San Diego (CA), see www.josre.org.

Karuppannan, S. and Sivam, A. (2009) Sustainable Development and Housing Affordability. Paper presented at the ENHR Conference, Prague, 29 June – 1 July, 2009.

Kauko, T. (2001) Combining Theoretical Approaches: The Case of Urban Land Value and Housing Market Dynamics, Review Article. *Housing, Theory and Society*, 18(3/4), pp. 167–173.

Kauko, T. (2002) Modelling Locational Determinants of House Prices: Neural Network and Value Tree Approaches. Doctoral dissertation, Utrecht, 2002, 251 p. http://igitur-archive.library.uu.nl/dissertations/2002-1204-091756/inhoud.htm.

Kauko, T. (2003) Planning Processes, Development Potential and House Prices: Contesting Positive and Normative Argumentation, Focus Article. *Housing, Theory and Society*, 20(3), pp. 113–126.

Kauko, T. (2004a) Sign Value, Topophilia and House Prices. *Environment and Planning A*, 36(5), pp. 859–878.

Kauko, T. (2004b) Infusing 'Institution' and 'Agency' into House Price Analysis. *Urban Studies*, 41(8), pp. 1507–1519.

Kauko, T. (2004c) Towards the 4th Generation – An Essay on Innovations in Residential Property Value Modelling Expertise. *Journal of Property Research*, 21(1), pp. 75–97.

Kauko, T. (2005) *Comparing Spatial Features of Urban Housing Markets: Recent Evidence of Submarket Formation in Metropolitan Helsinki and Amsterdam.* DUP Science Publication, Delft.

Kauko, T. (2006a) *Urban Housing Patterns in a Tide of Change: Spatial Structure and Residential Property Values in Budapest in a Comparative Perspective.* DUP Science Publication, Delft.

Kauko, T. (2006b) What Makes a Location Attractive for the Housing Consumer? Preliminary Findings from Metropolitan Helsinki and Randstad Holland Using the Analytical Hierarchy Process. *Journal of Housing and the Built Environment*, 21, pp. 159–176.

Kauko, T. (2008a) From Modelling Tools towards the Market Itself – An Opportunity for Sustainability Assessment? *International Journal of Strategic Property Management*, 12, pp. 95–107.

Kauko, T. (2008b) AVMs, Empirical Modelling of Value, and Systems for Market Analysis. In: Kauko, T. and d'Amato, M. (Eds.): *Mass Appraisal Methods – An International Perspective for Property Valuers*. RICS Series, Wiley-Blackwell, Oxford, pp. 307–319.

Kauko, T. (2009a) Policy Impact and House Price Development at the Neighbourhood-Level – A Comparison of Four Urban Regeneration Areas Using the Concept of 'Artificial' Value Creation. *European Planning Studies*, 17(1), pp. 85–107.

Kauko, T. (2009b) Classification of Residential Areas in the Three Largest Dutch Cities, Using Multidimensional Data. *Urban Studies*, 46(8), pp. 1639–1663.

Kauko, T. (2009c) The Housing Market Dynamics of two Budapest Neighbourhoods. *Housing Studies*, 24(5), pp. 587–610.

Kauko, T. (2010) Value Stability in Local Real Estate Markets. *International Journal of Strategic Property Management*, 14, pp. 191–199.

Kauko, T. (2011a) An Evaluation of the Sustainability of Inner City Residential Projects. *Housing, Theory and Society*, 28(2), pp. 144–165.

Kauko, T. (2011b) On Sustainable Urban Property Development – The Case of Hungary. Presented at the third *ReCapNet* Conference: "Real Estate Markets and Capital Markets" *of ZEW*, Mannheim, Germany, 14–15 October.

Kauko, T. (2013) Self-Organizing Map Algorithms to Identify Sustainable Neighbourhoods with an Example of Szeged (Hungary). Paper presented at the symposium "GIS Ostrava 2013 — Geoinformatics for City Transformation", January 21–23, Ostrava.

Kauko, T. and d'Amato, M. (2008) Introduction: Suitability Issues in Mass Appraisal Methodology. In: Kauko, T. and d'Amato, M. (Eds.): *Mass Appraisal Methods – an International Perspective for Property Valuers*. RICS Series, Wiley-Blackwell, Oxford, pp. 1–19.

Keen, S. (2009) Mad, Bad, and Dangerous to Know. *Real-World Economics Review*, 49, 12 March 2009, pp. 2–7. http://www.paecon.net/PAEReview/issue49/Keen49.pdf

Keivani, R. (2009) A Review of the Main Challenges to Urban Sustainability. *International Journal of Urban Sustainable Development*, 1(1–2), pp. 5–16.

Keskin, B. (2008) Hedonic Analysis of Price in the Istanbul Housing Market. *International Journal of Strategic Property Management*, 12, pp. 125–138.

Kloosterman, R.C. and Lambregts, B. (2007) Between Accumulation and Concentration of Capital: Toward a Framework for Comparing Long-Term Trajectories of Urban Systems. *Urban Geography*, 28(1), pp. 54–73.

Kohonen, T. (1995) *Self-Organizing Maps*. Springer Series in Information Sciences, Springer-Verlag, Germany.

Kohonen, T., Hynninen, J., Kangas, J. and Laaksonen, J. (1996) SOM_PAK: The Self-Organizing Map Program Package. Helsinki University of Technology, Faculty of Information Technology, Laboratory of Computer and Information Science. Report A31.

Kongkajaroen, P., Panichpathom, S. and Ngarmyarn, A. (2014) The Attitude of Intention to Purchase Green Condominium by Generation Y Consumers. Paper presented at the European Real Estate Society (ERES) Conference, 25–28 June, 2014, Bucharest, Romania.

Kontokosta, C.E. (2011) Greening the Regulatory Landscape: The Spatial and Temporal Diffusion of Green Building Policies in U.S. Cities. *Journal of Sustainable Real Estate*, 3(1), pp. 68–90.

Koopman, M. (2008) *The Spatial Foundations of the Housing Market.* Unpublished manuscript.

Kovács, Z. (2009) Social and Economic Transformation of Historical Neighbourhoods in Budapest. *Tijdschrift voor Economische en Sociale Geografie*, 100(4), pp. 399–416.

Kryvobokov, M. (2004) Urban Land Zoning for Taxation Purposes in Ukraine. *Property Management*, 22(3), pp. 214–229.

Leishman, C. and Warren, F. (2010) Planning for Consumers' New-Build Housing Choices. In: Adams, D., Watkins, C. and White, M. (Eds.): *Planning, Public Policy & Property Markets.* Wiley-Blackwell, Oxford, pp. 167–184.

Lentz, G.H. and Wang, K. (1998) Residential Appraisal and the Lending Process: A Survey of Issues. *Journal of Real Estate Research*, 15, pp. 11–39.

Locsmándi, G. (2006) National report – Hungary and Budapest.

Lorenz, D., d'Amato, M., Des Rosiers, F., Elder, B., van Genne, F., Hartenberger, U., Hill, S., Jones, K., Kauko, T., Kimmet, P., Lorch, R., Lutzkendorf, T. and Percy, J. (2008) *Sustainable Property Investment & Management – Key Issues & Major Challenges*, RICS. http://www.rics.org/site/download_feed.aspx?fileID=5 227&fileExtension=PDF [accessed 22 February, 2011].

Lorenz, D., Trück, S. and Lützkendorf, T. (2007) Exploring the Relationship between the Sustainability of Construction and Market Value: Theoretical Basics and Initial Empirical Results from the Residential Property Sector. *Property Management*, 25(2), pp.119–149.

Lucas, K, Halden, D. and Wixey, S. (2010) Transport Planning for Sustainable Communities. In: Manzi, T., Lucas, K., Lloyd-Jones, T. and Allen, J. (Eds.): *Social Sustainability in Urban Areas. Communities, Connectivity and the Urban Fabric.* Earthscan, London and Washington, DC, pp. 121–140.

LUDEN (2012) *Towards a New Paradigm for Local Urban Development.* http://qeceran.cluster003.ovh.net/documents/Towards-a-new-paradigm-for-Local-Urban-Development.pdf [accessed 24 November, 2012].

Lützkendorf, T., Fan, W. and Lorenz, D. (2011) Engaging Financial Stakeholders: Opportunities for a Sustainable Built Environment. *Building Research & Information*, 39(59), pp. 483–503.

Lützkendorf, T. and Lorenz, D. (2007) "Green Buildings" – Just Environmentally Sound or also Economical and a Stable Investment? In: Verband Deutcher Pfandbriefbanken (Ed.): *Real Estate Banking 2007–2008*, 6th ed. VBF, Association of German Pfandbrief Bank, Berlin, pp. 58–68.

Lützkendorf, T. and Lorenz, D. (2014) Sustainability Metrics – Translation and Impact on Property Investment and Management. A report by the property working group of United Nations environment programme finance initiative, May, UNEP, Brussels.

Lux, M., Sunega, P., Mikeszová, M. and Kostelecky, T. (2008) *Housing Standards 2007/2008: The Factors Behind the High Prices of Owner-Occupied Housing in Prague.* Institute of Sociology, Prague.

MacDonald, H. (1996) The Rise of Mortgage-Backed Securities: Struggles to Reshape Access to Credit in the USA. *Environment and Planning A*, 28(7), pp. 1179–1198.

Mace, A., Hall, P. and Gallent, N. (2007) New East Manchester: Urban Renaissance or Urban Opportunism? *European Planning Studies*, 15(1), pp. 51–65.

Macintosh, A. (2010) Visions of Tomorrow to Plan our Lives Today – Sustainable Development and the Real Estate Industry. *EU Sustainable Energy Week*, 22–26 March, 2010, Brussels.

Maclennan, D. (1977) Some Thoughts on the Nature and Purpose of House Price Studies. *Urban Studies*, 14, pp. 59–71.

Maclennan, D. and Tu, Y. (1996) Economic Perspectives on the Structure of Local Housing Systems. *Housing Studies*, 11(3), pp. 387–406.

Manzi, T., Lucas, K., Lloyd-Jones, T. and Allen, J. (Eds.) (2010a) *Social Sustainability in Urban Areas. Communities, Connectivity and the Urban Fabric*. Earthscan, London and Washington, DC.

Manzi, T., Lucas, K., Lloyd-Jones, T. and Allen, J. (2010b) Understanding Social Sustainability: Key Concepts and Developments in Theory and Practice. In: Manzi, T., Lucas, K., Lloyd-Jones, T. and Allen, J. (Eds.): *Social Sustainability in Urban Areas*. Earthscan, London and Washington, DC, pp. 1–28.

Martin, R. and Sunley, P. (2006) Path Dependence and Regional Economic Evolution. *Journal of Economic Geography*, 6, pp. 395–437.

Massimo, D.E. (2011) Emerging Issues in Real Estate Appraisal: Market Premium for Building Sustainability. Presented at the 41th meeting of *CeSET* – Italian Association of Appraisers and Land Economists, Rome, Italy, 14–15 November, 2011.

McMillen, S. (2005) *Better Designed Buildings: Improving the Valuation of Intangibles*. Eclipse Research Consultants Cambridge.

Medalen, T. (2006) Fortetting = Forfall (in Norwegian)? *Adresseavisen*, Kronikk, 13 November.

Meen, D. and Meen, G. (2003) Social Behaviour as a Basis for Modelling the Urban Housing Market: A Review. *Urban Studies*, 40(5–6), pp. 917–935.

Mehdizadeh, R., Fischer, M. and Celoza, A. (2013) LEED and Energy Efficiency: Do Owners Game the System? *Journal of Sustainable Real Estate*, 5(1), pp. 23–34.

Munday, M. and Roberts, A. (2006) Developing Approaches to Measuring and Monitoring Sustainable Development in Wales: A Review. *Regional Studies*, 40(5), pp. 535–554.

Musterd, S. and Deurloo, R. (2005) Amsterdam and the Preconditions for a Creative Knowledge City. *Tijdschrift voor Economische en Sociale Geografie*, 97(1), pp. 80–94.

Openshaw, S., Blake, M. and Wymer, C. (1994) Using Neurocomputing Methods to Classify Britain's Residential Areas. Working Paper 94/17, School of Geography, University of Leeds.

Op't Veld, D., Bijlsma, E. and van de Hoef, P. (2008) Automated Valuation in the Dutch Housing Market: The Web-Application 'MarktPositie' Used by NVM-Realtors. In: Kauko, T. and d'Amato, M. (Eds.): *Advances in Mass Appraisal Methods*. Wiley-Blackwell, Oxford, pp. 70–90.

Percy, S. (2003) New Agendas. In: Couch, C., Fraser, C. and Percy, S. (Eds.): *Urban Regeneration in Europe*. Real Estate Issues, Wiley-Blackwell, Oxford, pp. 200–209.

Pareja Eastaway, M. and Støa, E. (2004) Dimensions of Housing and Urban Sustainability. *Journal of Housing and the Built Environment*, 19, pp. 1–5.

Perspectives (2013) Making Cities Smart, pp. 12–19. (A publication by Intergraph).

Prasad, N. and Richards, A. (2008) Improving Median Housing Price Indexes through Stratification. *Journal of Real Estate Research*, 30(1), pp. 45–71.

Raslanas, S., Zavadskas, E.K. and Kaklauskas, A. (2010) Land Value Tax in the Context of Sustainable Urban Development and Assessment. Part I – Policy Analysis and Conceptual Model for the Taxation System on Real Property. *International Journal of Strategic Property Management*, 14, pp. 73–86.

Ratcliffe, J., McIntosh, A. and Brown, S. (2010) *Built Environment Foresight 2030: the sustainable development imperative*. RICS Foundation, The University of Salford, Futures Academy and King Sturge, Salford.

Rauterkus, S.Y. and Miller, N.G. (2011) Residential Land Values and Walkability. *Journal of Sustainable Real Estate*, 3(1), pp. 23–43.

Reese, L. and Sands, G. (2007) Sustainability and Local Economic Development in Canada and the United States. *International Journal of Sustainable Development Planning*, 2(1), pp. 25–43.

Regjeringen (2011) *Framtidens byer* [in Norwegian]. http://www.regjeringen.no/nb/sub/framtidensbyer/om-framtidens-byer.html?id=548028 [accessed 2 July, 2011].

The Renewable Energy & Energy Efficiency Partnership (REEEP) (2011) http://www.reeep.org/ [accessed 13 February, 2011].

Ribeiro, F.L. (2008) Urban Regeneration Economics: The Case of Lisbon's Old Downtown. *International Journal of Strategic Property Management*, 12, pp. 203–213.

RICS (2007) Financing and Valuing Sustainable Property: We Need to Talk. Findings in Built and Rural Environments (FiBRE), RICS Research, April. www.rics.org.

RICS (2008) Urban Regeneration Is in Your Hands! Findings in Built and Rural Environments (FIBRE), RICS EU Public Affairs, Brussels, September.

RICS (2010) Energy Efficiency and Value Project. Final report. RICS Communities, March.

RICS (2011) A Vision for Sustainability. RICS Corporate Responsibility report. August 2010 – July 2011.

Runde, T. and Thoyre, S. (2010) Integrating Sustainability and Green Building into the Appraisal Process. *Journal of Sustainable Real Estate*, 2(1), pp. 221–248.

Sager, T. (2009) Planners' Role: Torn between Dialogical Ideals and Neo-Liberal Realities. *European Planning Studies*, 17(1), pp. 65–84.

Sager, T. (2010) Role Conflict: Planners Torn between Dialogical Ideals and Neo-Liberal Realities. In: Hillier, J. and Healey, P. (Eds.): *The Ashgate Research Companion to Planning Theory: Conceptual Challenges for Spatial Planning*, Ashgate, Farnham, pp. 183–214.

Sapir, J. (2008) Global Finance in Crisis: A Provisional Account of the "Subprime" Crisis and How We Got into It. *Real-World Economics Review*, 46, 20 May 2008, pp. 82–101. http://www.paecon.net/PAEReview/issue46/Sapir46.pdf.

Sayce, S., Ellison, L. and Parnell, P. (2007) Understanding Investment Drivers for UK Sustainable Property. *Building Research & Information*, 35(6), pp. 629–643.

Sayce, S., Sundberg, A. and Clements, B. (2010) Is Sustainability Reflected in Commercial Property Prices: An Analysis of the Evidence Base. RICS Research, January.

Schumann, B. (2010) Impact of Sustainability on Property Values. Master's Thesis, International Business School, University of Regensburg.

Schwegler, B. (2006) Entrepreneurial Governance and the Urban Restructuring of a Slovakian Town. In: Tsenkova, S. and Nedović-Budić, Z. (Eds.): *The Urban Mosaic of Post-Socialist Europe. Space, Institutions and Policy.* Physica-Verlag, Heidelberg, pp. 295–318.

Seyfang, G. (2006) Sustainable Consumption, the New Economics and Community Currencies: Developing New Institutions for Environmental Governance. *Regional Studies*, 40(7), pp. 781–791.

Shanmugaratnam, N. (1990) Sustainable Development – A Conceptual Overview. Proceedings of The Symposium on the Sustainability of Agricultural Production Systems in Sub-Saharan Africa, September 4–7, 1989, Aas, Norway, NorAgric, Occasional Papers Series C, pp. 25–42.

Silhankova, V. and Pondelicek, M. (2010) Evaluation of Sustainable Regional Land Use. Paper presented at the EURA Conference 'Understanding City Dynamics', 24–26. September, 2010, Darmstadt, Germany.

Smets, P. and Kreuk, N. (2008) Together or Separate in the Neighbourhood? Contacts between Natives and Turks in Amsterdam. *The Open Urban Studies Journal*, 1, pp. 35–47.

Söderbaum, P. (2009) A Financial Crisis on Top of the Ecological Crisis: Ending the Monopoly of Neoclassical Economics. *Real-World Economics Review*, 49, 12 March 2009, pp. 8–19. http://www.paecon.net/PAEReview/issue49/Soderbaum49.pdf.

Spaargaren, G. (2003) Sustainable Consumption: A Theoretical and Environmental Policy Perspective. *Society and Natural Resources*, 16, pp. 687–701.

Spielman, S.E. and Thiel, J-C. (2007) Social Area Analysis, Data Mining, and GIS. *Computers, Environment and Urban Systems*, 32, pp. 110–122.

Stenberg, J. (2008) Multidimensional Evaluation for Sustainable Development: Managing the Intermix of Mind, Artefact, Institution and Nature. In: Miller, D., Khakee, A., Hull, A. and Woltjer, J. (Eds.): *New Principles in Planning Evaluation*, Ashgate, Aldershot, pp. 35–53.

Stewart, D., Sirr, L. and Kelly, R. (2006) Smart Growth: A Buffer Zone between Decentrist and Centrist Theory? *International Journal of Sustainable Development Planning*, 1(1), pp. 1–13.

Støa, E. (2009) Housing in the Sustainable City – Issues for an Integrated Approach. In: Holt-Jensen, A. and Pollock, E. (Eds.): *Urban Sustainability and Governance*, Nova Science, NY, pp. 31–48.

Talen, E. (2011) Sprawl Retrofit: Sustainable Urban form in Unsustainable Places. *Environment and Planning B*, 38, pp. 952–978.

Taşan-Kok, T. (2010) Entrepreneurial Governance: Challenges of Large-Scale Property-Led Urban Regeneration Projects. *Tijdschrift voor Economische en Sociale Geografie*, 101(2), pp. 126–149.

Taylor Wessing (2009) Behind the Green Façade: Is the UK Development Industry Really Embracing Sustainability? Taylor Wessing sustainability report.

Thiet, V. (2011) Head of Development, ICADE Germany, (A Real Estate Development and Investment Company): "Cognitive Buildings". CognitiveCities conference, Berlin 26 February, 2011. http://conference.cognitivecities.com/speakers/#vini-tiet [accessed 4 July, 2011].

Trondheim Kommune (2010) *Kommuneplanens samfunnsdel 2009–2020* [in Norwegian]. http://www.trondheim.kommune.no/attachment.ap?id=33138 [accessed 2010].

Trondheim Kommune (2012) http://www.trondheim.kommune.no [accessed 11 May, 2012]

Tsenkova, S. (2006) The Post-Socialist Urban World (with Nedović-Budić, Z.). In: Tsenkova, S. and Nedović-Budić, Z. (Eds.): *The Urban Mosaic of Post-Socialist Europe. Space, Institutions and Policy*. Physica-Verlag, Heidelberg, pp. 349–366.

Turcu, C. (2009) In the Quest of Sustainable Communities: A Theoretical Framework to Assess the Impact of Urban Regeneration on Community Sustainability. In: Tsenkova, S. (Ed.), *Planning Strategies for Sustainable Cities*. University of Calgary, Calgary, pp. 37–66.

Turok, I. (2004) Cities, Regions and Competitiveness. *Regional Studies*, 38(9), pp. 1069–1083.

Uitermark, J., Duyvendak, J.W. and Kleinhans, R. (2007) Gentrification as a Governmental Strategy: Social Control and Social Cohesion in Hoogvliet, Rotterdam. *Environment and Planning A*, 39, pp. 125–141.

UN (2010) *Policy Framework for Sustainable Real Estate Markets. Principles and Guidance for the Development of a County's Real Estate Sector*. United Nations Economic Commission for Europe (UNECE), Working Party on Land Management (WPLA), Real Estate Market Advisory Group (REM), United Nations, Geneva.

Van den Berg, H. (2014) *Economic Growth and Development*, 2nd ed. World Scientific Publishing Co., Singapore.

Van der Maaten, E. (2010) Uncertainty, Real Option Valuation, and Policies toward a Sustainable Built Environment. *Journal of Sustainable Real Estate*, 2(1), pp. 161–181.

Varvarigos, D. and Zakaria, I.Z. (2011) Growth and Demographic Change: Do Environmental Factors Matter.

Vatn, A. (2005) *Institutions and the Environment*. Edward Elgar, Cheltenham.

Wagner, A., Gossauer, E., Moosmann, C., Gropp, T. and Leonhart, R. (2007) Thermal Comfort and Workplace Occupant Satisfaction – Results of Field Studies in German Low Energy Office Buildings. *Energy and Buildings*, 39, pp. 758–769.

Wallace, A. (2008) Knowing the Market? Understanding and Performing York's Housing. *Housing Studies*, 23(2), pp. 253–270.

Wallace, A., Jones, A. and Duffy, S. (2009) *Rapid Evidence Assessment of the Economic and Social Consequences of Worsening Housing Affordability*. The University of York, Centre for Housing Policy, York.

Wallner, H.P., Narodoslawsky, M. and Moser, F. (1996) Islands of Sustainability: A Bottom-Up Approach Towards Sustainable Development. *Environment and Planning A*, 28(19), pp. 1763–1778.

Warren-Myers, G. (2012) The Value of Sustainability in Real Estate: A Review from a Valuation Perspective. *Journal of Property Investment and Finance*, 30(2), pp. 115–144.

Warren-Myers, G. and Reed, R. (2010) The Challenges of Identifying and Examining Links between Sustainability and Value: Evidence from Australia and New Zealand. *Journal of Sustainable Real Estate*, 2(1), pp. 201–220.

Watkins, C.A. (2001) The Definition and Identification of Housing Submarkets. *Environment and Planning A*, 33, pp. 2235–2253.

Webber, R. (2007) The Metropolitan Habitus: Its Manifestations, Locations, and Consumption Profiles. *Environment and Planning A*, 39, pp. 182–207.

Wessel, T. (2002) Fra leie til eie – konvertering av leiegårder i norske byer (in Norwegian). *Tidsskrift for samfunnsforskning*, 43(3), pp. 299–331.

Wheeler, S.M. and Beatley, T. (2014) *The Sustainable Urban Development Reader*, 3rd ed. Routledge, Abingdon and New York.

White, P., Christodoulou, G., Mackettt, R., Titheridge, H., Thoreau, R. and Polak, J. (2010) In: Manzi, T. Lucas, K., Lloyd-Jones, T. and Allen, J. (Eds.) The Impacts of Teleworking on Sustainability and Travel. *Social Sustainability in Urban Areas*. Earthscan, London and Washington, DC, pp. 141–154.

Wilkinson, S.J., Van Der Kallen, P. and Kuan, L.P. (2013) The Relationship between the Occupation of Residential Green Buildings and Pro-Environmental Behavior and Beliefs. *Journal of Sustainable Real Estate*, 5(1), pp. 1–22.

Williams, K. (2010) Sustainable Cities: Research and Practice Challenges. *International Journal of Urban Sustainable Development*, 1(1–2), pp. 128–132.

Woon Amsterdam (in Dutch) (2007) Dienst Belastingen Gemeente Amsterdam & Makelaarsverening Amsterdam.

Young, S.T. and Dhanda, K.K. (2013) *Sustainability – Essential for Business*. Sage, London.

Yu, Sh.-M. and Tu, Y. (2011) Are Green Buildings Worth More Because They Cost More? IRES Working Paper Series IRES2011–023, August.

Zalejska-Jonsson, A. (2013) Impact of Energy and Environmental Factors in the Decision to Purchase or Rent and Apartment: The Case of Sweden. *Journal of Sustainable Real Estate*, 5(1), pp. 66–85.

Zietz, E.N. and Sirmans, G.S. (2004) An Exploration of Inner-City Property Markets. *Journal of Real Estate Literature*, 12(3), pp. 323–360.

Zuindeau, B. (2006) Spatial Approach to Sustainable Development: Challenges of Equity and Efficacy. *Regional Studies*, 40(5), pp. 459–470.

Interviews

Manuel Aalbers, Researcher, The University of Amsterdam, 25 July, 2002.

Axel Baudouin, Associate Professor, NTNU, Discussion, March 2007.

Zsuzsanna Földi, Urban Development Expert, Terra Studio Kft., Consultant, 27 April, 2005.

Gábor Locsmándi, Associate Professor, Department of Urban Studies at Budapest University of Technology and Economics, Budapest, 7 January, 17 February, 18 March and 25 June, 2004.

Csilla Sárkány, Manager (and György Alföldi, Member of the Board), Rév8, land management company of the local government authority in district VIII, 14 April, 2005.

Reynt-Jan Sluys formerly Researcher at OTB Research Institute of Housing and Mobility Studies, Delft University of Technology, Discussions, 2002 and 2003.

Gábor Soóki-Tóth, Managing Director, ECORYS Hungary, Real Estate Consultancy, Budapest, 15 April, 2005.

Árpád Szabó, Planning Consultant, 3 May, 2005.

Willem Teune, The Housing Office of the City of Amsterdam, SWD, 25 July, 2002.

Dávid Valkó, Otthon Center, Real Estate Agent, Leading Adviser, 3 May, 2005.

Andre van den Berg (2006) Officer, Oud-zuid, de Pijp, 1 February, Amsterdam.

Index

Note: figures and tables are denoted with italicized page numbers.

Taylor & Francis eBooks

Helping you to choose the right eBooks for your Library

Add Routledge titles to your library's digital collection today. Taylor and Francis ebooks contains over 50,000 titles in the Humanities, Social Sciences, Behavioural Sciences, Built Environment and Law.

Choose from a range of subject packages or create your own!

Benefits for you

» Free MARC records
» COUNTER-compliant usage statistics
» Flexible purchase and pricing options
» All titles DRM-free.

Benefits for your user

» Off-site, anytime access via Athens or referring URL
» Print or copy pages or chapters
» Full content search
» Bookmark, highlight and annotate text
» Access to thousands of pages of quality research at the click of a button.

REQUEST YOUR **FREE** INSTITUTIONAL TRIAL TODAY

Free Trials Available
We offer free trials to qualifying academic, corporate and government customers.

eCollections – Choose from over 30 subject eCollections, including:

Archaeology	Language Learning
Architecture	Law
Asian Studies	Literature
Business & Management	Media & Communication
Classical Studies	Middle East Studies
Construction	Music
Creative & Media Arts	Philosophy
Criminology & Criminal Justice	Planning
Economics	Politics
Education	Psychology & Mental Health
Energy	Religion
Engineering	Security
English Language & Linguistics	Social Work
Environment & Sustainability	Sociology
Geography	Sport
Health Studies	Theatre & Performance
History	Tourism, Hospitality & Events

For more information, pricing enquiries or to order a free trial, please contact your local sales team:
www.tandfebooks.com/page/sales

Printed in the United States
By Bookmasters